SATELLITE
INFORMATION
SYSTEMS

SATELLITE INFORMATION SYSTEMS

BY EDWARD BINKOWSKI

G.K. Hall Publishers ● Boston, Massachusetts

SATELLITE INFORMATION SYSTEMS

EDWARD BINKOWSKI

Copyright 1988
by G.K. Hall & Co.
70 Lincoln Street
Boston, Massachusetts 02111

Book production by Patricia D'Agostino

Copyediting supervised by Barbara Sutton

Library of Congress Cataloging-in-Publication Date

Binkowski, Edward S.
     Satellite information systems.

     (Professional librarian series)
     Includes index.
     1. Artificial satellites in telecommunication.
2. Information networks.   I. Title.   II. Series.
TK5104.B532   1988   384.3   88-16455
ISBN 0-8161-1856-6
ISBN 0-8161-1880-9 (pbk.)

To Alison

CONTENTS

Prologue
A DAY IN THE LIFE OF A
SATELLITE USER

In his Marin County, California, home, John Parker's morning routine is to scan the *Wall Street Journal* while listening to the weather report. This morning, after his usual check of commodities prices, he reads an article on arms control. Although he gives it no thought at all, the weather report is based on data from a GOES meteorological satellite; the *Wall Street Journal* is beamed to the West Coast by satellite; the information for the arms control article comes from a military reconaissance satellite; and the surprising move in commodities prices is based on information from a remote sensing satellite.

He works as a broker for a maritime insurance firm in San Francisco, which he hardly thinks of as a satellite business. This is despite the dozens of domestic and international calls he makes a day that go out over satellite, including those on his mobile phone as well as signals for a beeper on field trips. The bank transfers his firm authorizes, the global scheduling activities it carries on, and the navigation of the ships it insures are all performed over satellite systems. The unsung salvation of the firm on several occasions has been the SARSAT search-and-rescue satellite system, which has prevented several losses at sea.

His wife, Ann, works as a policy analyst in a corporate public affairs office. Her package of computer printout in the morning prompts a video-conference at lunchtime, the results of which are announced that afternoon to all corporate locations by electronic mail. The package, the conference, and the announcement all travel over satellite systems. The data bank she references and the shared library resources she uses are made possible in the same way.

Their daughter, Jenny, is a college student and has no connection with satellites at all except, of course, for the classes she takes over her university's satellite system.

The three gather in the evening for the American tradition of television. After the news, with much of the content coming from satellites, they argue over whether to watch the latest HBO blockbuster or a network situation comedy. Whichever they choose, the broadcast will be transmitted by satellites—as will the radio program the loser will dial in instead.

In hundreds of ways, large and small, satellite information systems touch our lives every day. This book is a brief guide to how they have evolved, how they operate today, and how they will affect us even more tomorrow.

Part 1
What Are Satellite Information Systems?

Chapter 1
Introduction

DEFINITION

Satellite information systems are <u>interconnected</u> systems of artificial satellites in earth's orbit that gather, receive, process, and transmit data and communication for use on the surface. This book is intended to increase the understanding of these systems for the nontechnical user with respect to three threshold questions:

- What are the costs and benefits of *satellite* systems relative to other competing modalities?
- How are different types of *information* best manipulated and distributed?
- How are satellite systems best integrated with other information *systems*?

Typical readers will not normally be faced with the task of constructing and operating a satellite information system from scratch. As the prologue makes clear, however, in the course of a normal business day they will be direct or indirect users of such systems many times over. For such users in business, government, and education—whether they are part of the so-called information industry (e.g., broadcasting, publishing, finance, telecommunications, data processing, library management, etc.) or simply dependent on effective information distribution for carrying on their activities—this book provides a description of the choices open to them and, more often, the choices that have been made for them.

Finally, the discussion here is directed to the practical (and usually commercial) interests of the users, and is not overly concerned with national security, global diplomacy, or the latest engineering advances. Nevertheless, until at least 1984, straightforward commercial interests played at best a minor role in the development of satellite information systems compared to military, political, and technological pressures and constraints. The remainder of this chapter (a brief history of the satellite age and an even briefer technical background) attempts to provide this broader context, and show how satellite information systems moved, in two generations, from a science-fiction writer's dream to carrying more than two-thirds of world telecommunications.

A BRIEF HISTORY

The 1940s

The first reference to satellite information systems is traditionally credited to a proposal published in October 1945 by Arthur C. Clarke, later to become known as a science-fiction writer.[1] The Clarke article, originally ambitiously titled "The Future of World Telecommunications" and later, more modestly, "Extraterrestrial Relay," suggests the creation of a global communications system that could provide coverage of any point on earth with just three satellites, without any need for cables or relays.

The power of the Clarke proposal lies in his identification of the *geostationary orbit* for a satellite. The ability of any satellite to stay in orbit is a function of its altitude and speed. If a satellite moves too slowly, earth's gravitational forces will rapidly drag it down; if it moves too quickly it will escape earth's gravitational field altogether and fly off into deep space. For any given altitude, there is a critical speed that must be maintained to avoid both events. If that speed were matched by the angular velocity of the earth's rotation, a satellite traveling in the same rotational direction would always appear to be over the same spot on the earth's surface, that is, to be geostationary.

The altitude for this critical geostationary speed is approximately 22,300 miles. A satellite in orbit around the earth's equator at this height will be in view of more than 40 percent of the world and, thus, three satellites could easily provide coverage from any point on earth to any other point. For any satellite system not in geostationary orbit, any given earth location would always be "losing sight of" one particular satellite and would require a much larger system of satellites in orbit in order not to lose communications capacity.

Clarke's proposal was not taken very seriously at the time, even by the author himself. In later years he confessed, with some chagrin, to having sold one of the most commercially viable ideas of the twentieth century for $40 in "A Short Pre-History of Comsats, or How I Lost a Billion Dollars in My Spare Time."[2] Brilliant though his insight may have been, it was not patentable. Furthermore, the 22,300-mile requirement seemed technically infeasible in 1945, especially given Clarke's belief that his satellite system would have to be manned in order to be controlled successfully. Finally, Clarke wrote his article from his position as Secretary of the British Interplanetary Society, a group of rocket research and science-fiction enthusiasts.

The British Interplanetary Society, however, had its counterpart in Germany in the Rocket Travel Society (*Verein für Raumschiffart*). Forbidden to engage in artillery research by the Treaty of Versailles, the German government actively supported the amateur group and, in 1933 approached then 22-year-old Wernher von Braun to construct a new experimental military rocket.[3] Ultimately, the "rocket team" developed the V-1 and V-2 weapons of World War II.

Immediately after their capture by American forces in 1945, the German rocket scientists told of plans for artificial earth satellites and space stations. A report to the United States government was submitted in August 1945, and by October, that is, by the time of publication of Clarke's paper, the Navy Bureau had set up an Earth Satellite Vehicle Program.[5]

The Navy's chief interest was in the use of satellites as navigational aids; the Navy approached the Air Force for a joint development program, but the Air Force demurred, claiming satellites had no military value. At the same time (May 1946), a private study commissioned by the Air Force and conducted by RAND (then a proprietary organization of the government) found significant communications, reconnaissance, and scientific uses for satellites.[6] Naturally enough, von Braun was working for neither the Navy nor the Air Force but the Army, whose interest remained restricted to missile development (although the Army Signal Corps did bounce a radar signal off the moon in January 1946 as a technical challenge in Project Diana[7]).

By the end of 1948, interservice rivalries, lack of immediate progress, failure to recognize military applications, dramatically slashed defense budgets as part of the Department of Defense consolidation, and disinterest of the new Truman administration all worked together to terminate the scarcely begun satellite research program.

The 1950s

Other than in the minds of RAND researchers, satellite research remained static until 1954, when President Eisenhower established the Technological Capabilities Panel to advise on the technological possibilities for forestalling a surprise Soviet attack.[8] This committee also drew its motivation from intelligence failures of the Korean War, continuing disarmament talks, and emerging appreciation of the Soviet theft of atomic bomb secrets. Headed by Dr. James Killian, president of Massachusetts Institute of Technology, this committee strongly recommended active aerial surveillance of the Soviet Union by "spy planes" in what was to become the U-2 program. Recognizing the legal and technical vulnerability of this activity, the group also recommended immediate development of a satellite intelligence-gathering capability.[9]

Concerned that even a satellite in earth orbit passing over another country might be construed as violation of sovereign air space,[10] Eisenhower wanted the satellite program to appear to be nonmilitary and to set a precedent for satellite overflight. Following a RAND recommendation, an announcement was made in April 1955 that the United States would launch a satellite to perform scientific research as part of the International Geophysical Year, a cooperative international research program sponsored by the United Nations through the International Council of Scientific Unions and actually running from July 1957 through December 1958.[11]

Use of satellites for telecommunications was still not being actively considered. In 1952 Dr. John Pierce, of Bell Telephone Laboratories, wrote a paper discussing the improved technical feasibility of Clarke's proposal; however, this paper was written under a pseudonym for the magazine *Amazing Science Fiction*.[12] In 1954, spurred by advances in microwave and transistor research, Pierce went public with his conclusions, but he received a largely skeptical response.[13] Commercial communications satellite systems were still more than a decade off.

The government satellite program received proposals from the Air Force, based on its Atlas missile; the Army, with its Redstone missile developed by von Braun; and the Navy, without an existing launch vehicle but with an institutional history of interest in satellites and of use of rockets for atmospheric research rather than missiles. The civilian character of the Eisenhower directive led to the choice of the Navy as the winner, to the considerable regret of the Army and von Braun. What followed were two years of confusion, shoestring budgets, continuing interservice rivalries, contracting disputes, and near total unawareness that the Soviets were involved in a parallel effort.[14]

This unawareness was rudely shaken by the launch of Sputnik by the Soviet Union on 4 October 1957. The first launch by the Navy of the inauspiciously named Vanguard on December 6 was an ignominious failure, the rocket exploding four feet off the ground. Von Braun and the Army, chafing at the bit, had claimed immediately after Sputnik that they could make a successful launch within sixty days and, in fact, had been ready to do so for the past year. (Von Braun had even planned for a clandestine launch the year before.[15]) The Vanguard failure gave the go-ahead signal to the Army, and its Jupiter missile was able to put Explorer 1 into orbit on 31 January 1958.

Explorer 1 and its immediate successors, including the first successful Vanguard launch in March, were all dedicated to scientific measurement, Explorer 1 discovering the Van Allen belt. In December 1958 Project Score became the first communications satellite. The Score satellite could carry one voice channel or seven teletype channels, either in real time or by delayed recording, to four earth stations. The batteries powering the satellite failed after one month, but not before relaying a Christmas message from President Eisenhower around the world.[16] Finally, in February 1959 in Discoverer 1 the military had its prototype spy satellite capable of delivering photographic evidence to the ground, but through actual delivery of the film rather than telecommunications.

Even after the Sputnik upset Eisenhower insisted upon at least nominal civilian control of the national space program. He also created the post of the National Science Adviser, filled by Dr. Killian of the Killian report, whose first major recommendations led to the almost overnight creation of NASA, the National Aeronautics and Space Administration, in 1958. NASA's initial relationships with the military were cool at best, and sharp conflicts remain with the Air Force even today. NASA's immediate mission to salvage national prestige was to put a man in space, and once again satellite development was temporarily on hold.[17]

The 1960s

This decade began with a clear demonstration of the farsightedness of the Killian report: the shooting down over Soviet territory of a U-2 spy plane in May 1960. A temporary crisis in United States–Soviet relations ensued, including collapse of Eisenhower–Krushchev summit talks in Paris. Neither the East-West nor intra-United States government rivalries were aided by the early administration statements that the U-2 was a NASA weather plane. Nevertheless, later that same month the first successful MIDAS (Missile Defense Alarm System) satellite went into orbit, to be used to detect missile launches through infrared radiation sensing. Routine photographic flights followed that year.[18] In addition, 1960 saw the successful launch of the first of the Transit system of satellites to be used as navigational aids for Polaris submarines and the first earth observation and meteorological satellite, Tiros 1, both in April of that year.[19]

The year was nowhere near as encouraging for advances in communications satellites. Echo 1 was launched in August. Echo, nothing more than a large Mylar balloon covered with aluminum, was intended to serve as a passive reflector for radio and television signals without any processing or enhancement. The project was most notable because it marked the first collaboration with Bell Telephone Laboratories researchers. It was rapidly concluded that active repeater satellites (that is, satellites processing, amplifying, and redirecting the signals sent to them) were superior systems. The first beneficiary of this conclusion was Courier 1-B, the first component of a global military communications system, in October 1960, although Courier functioned in orbit for only three weeks.[20] Active repeater satellites were to be the rule thereafter.

Military surveillance satellites dominated 1961 advances, with development emphasis otherwise placed on the Project Mercury "first man in space" race. Especially successful was the Samos series, which took Polaroid pictures in space, digitized the pictures electronically, and radioed these versions of the pictures to waiting ground stations. Although not as spectacular as mid-air catches of film canisters from Discoverer satellites, the Samos system produced intelligence in enormously greater quality and quantity. One of the first discoveries of Samos 2 was the nonexistence of the much-ballyhooed missile gap of the 1960 presidential campaign.[21]

In 1961 ten Western European countries, led by France and Germany, formed the loosely knit European Launch Development Organization (ELDO) with the intent of creating a space program independent of both the United States and the Soviet Union. Although it was followed next year by the European Space Research Organization (ESRO), these early efforts faced too many political and technological obstacles and were completely unsuccessful.[22]

After John Glenn's successful orbital flight in February 1962, attention could be turned to communications satellites once more, and in July, Telstar was placed into orbit. Designed and built by American Telephone and Telegraph (AT&T) as a joint effort with NASA, Telstar was the first satellite built with private funds, was the first active relay satellite for public communications, and provided the first live television broadacst between continents. Telstar was also a proving ground for the first computer-to-computer satellite communications. By NASA's own enabling legislation, Telstar was government property, but NASA waived its right in favor of AT&T. Ironically, Telstar was rendered inoperable by radiation from American high-altitude nuclear tests six months later.[23]

Although Telstar competitors proposed by RCA and Western Union were waiting in the wings, within a week of the launch President Kennedy proposed the formation of the Communications Satellite Corporation (ComSat), a quasi-private entity formed of the major United States common carriers to develop and administer domestic satellite communications.[24]

The first direct attempts at placing a satellite into geostationary orbit were made in 1963. Syncom I, launched in February, was a mechanical failure. Syncom II, launched in July, achieved what was euphemistically called a near stationary orbit. Transmissions were made to ships at sea, and television signals were relayed (but without audio) from locations wavering north and south of the equator. Finally—nineteen years after Clarke's proposal, seven years after Sputnik—Syncom III was placed in geostationary orbit in August 1964 just in time for coverage of the Tokyo Olympics.[25]

As with Telstar, competitors to and extensions of Syncom III were near actuality. Almost simultaneously with its launch, however, the International Telecommunications Satellite Consortium, Intelstat, was formed as the international analog to ComSat and with ComSat as its original manager. Intelsat's charter was the provision of international satellite communications, with all government parties pledging not to allow competing systems. Hence, from the very start "commercial" communications were a monopolistic closed club. In June of 1965 Intelsat's first satellite, Early Bird, was put into geostationary orbit. Early Bird, just as Intelsat itself at that time, was American in all but name: manufactured by Hughes, launched by NASA, operated by ComSat, and carrying preponderantly United States communications. It was joined by neighbors over the next two years so that by 1967, twenty-two years after the Clarke proposal and ten years after Sputnik, worldwide coverage by a network of communications satellites in geostationary orbit was finally a reality.[26]

This global system does not include the Soviet Union, which with other Eastern bloc countries, formed its own Intersputnik system rather than accept an invitation to joint Intelsat. The first satellite in the Soviet system, Molniya 1, was also launched in 1965. Due to the difficulty of transmitting to extreme northern latitudes from an equatorial orbit, the Soviet system does not use geostationary satellites.[27]

After this rapid sequence of successes, communication satellite developments were yet again sidetracked, first by the Apollo missions for a manned moon landing, and then by the inevitable drastic budget cuts undermining NASA after actual completion of a program with no clear follow-up in mind. The military once more became the prime sponsor of communications research and the decade turned nearly full circle, beginning and ending with almost total lack of civilian and commercial concerns.

The 1970s

The early 1970s saw a continued decline in NASA's budget and prestige. In the private sector, the common carriers who were the principal actors in ComSat, uncomfortable with both the level and vagueness of government control, dropped out one by one until the entity was almost entirely publicly held. The year 1972 saw the reversal of this decline on several fronts. First, the Federal Communications Commission (FCC) ruled in its "Domsat," or "open skies," decision that private parties could compete with ComSat for the provision of domestic satellite communication services.[28] This decision signaled the beginning of a new space race where, this time, the competitors were United States corporations. In 1974 American Satellite Corporation, a new subsidiary of the Fairchild Corporation, won this race with its WESTAR satellites, becoming the first privately owned system nearly thirty years after the Clarke proposal. Then RCA followed with SATCOM in 1975, and Comstar was an AT&T-GTE joint venture in 1976.[29]

These corporations were not the only ones to see and seize opportunities in 1972. The Canadian government began its own domestic satellite communications system (launched, of course, by NASA) with ANIK 1.[30] Out of the ruins of the ELDO and ESRO programs, but now with the organized backing of the European Economic Community (i.e., the Common Market), the European Space Agency (ESA) was created, this time to enjoy substantial success.[31] The first nonmilitary, remote sensing satellites were launched by the United States in the LANDSAT series.[32]

Last but not least, early in the 1972 campaign year, President Nixon sought to create a parallel challenge to Kennedy's call for man on the moon by the end of the decade. His advisers came up with three proposals: a man on Mars by the end of the century; a manned space station; and a manned, reusable spacecraft. The budget and timing demands of the first two options quickly led to their dismissal. The third was acted upon and, although all but ignored in the unfolding Watergate saga that year, became the beginning of the space shuttle program.[33]

Intelsat had been operating for nine years on the basis of the interim 1964 agreements, during which time the United States' role, though still dominant, had consistently shrunk. In 1973 the permanent agreements were finally signed after much negotiation, which gave ComSat a more limited role in control of Intelsat (and by the end of the decade Intelsat management was to be determined by vote).[34] One of the immediate consequences of this shift was the decision to allow countries to operate their own domestic satellite communications using excess capacity on the Intelsat system. Taking advantage of this opportunity in the 1970s were Algeria, Brazil, Chile, Colombia, India, Malaysia, Mexico, Nigeria, Oman, Peru, the Philippines, Saudi Arabia, the Sudan, and Zaire. In 1976 Indonesia followed Canada in initiating its own national system, Palapa.[35]

Intelsat's new-found success led to the creation of a parallel system, Inmarsat, in 1978, which was to operate maritime rather than intercontinental satellite information systems, and this time the Soviet Union participated from the outset.[36]

The decade of the 1970s, begun at such a low point, ended with two events that marked the coming of age for satellite information systems. First, in October 1979 technology finally outstripped both the need for (and the ability of) regulation to control reception of satellite signals, as the FCC recognized in its ruling overturning licensing requirements for satellite earth stations.[37] Any party, anywhere, could build a backyard dish. Then on Christmas Eve, Arianespace, a quasi-public French corporation, successfully launched an ESA satellite into orbit, the first such nongovernmental launch undertaken by a country other than the two superpowers.[38]

The 1980s

The growth of satellite information systems continued to accelerate in the first half of the decade. In 1981 Satellite Business Systems (SBS), a joint venture of Aetna Life Insurance and IBM (ComSat, an earlier partner, dropped out), joined the group of American corporations operating domestic satellite networks, and First Interstate Bancorp became the first operator of a private information network for its own corporate use.[39] The biggest news of 1981, however, was the maiden flight of the space shuttle in April. The shuttle was planned to be the satellite delivery system of choice for NASA, and Arianespace immediately countered with an open offer to take on commercial satellite launches.[40]

In 1982 the FCC dissolved the last effective constraints on

ComSat by ruling that ComSat could initiate services in its own right and compete head to head with the common carriers who had been its original operators.[41] At the same time, applications for direct broadcasting satellites (i.e., satellites capable of providing television broadcasting directly into homes without any intermediate earth station processing) were invited by the FCC, and eight companies responded by the end of the year.[42]

In his State of the Union address in January 1983, President Reagan made a commitment to a manned space station by 1992. Momentous though this decision was, it was dwarfed by his Strategic Defense Initiative, or "Star Wars," proposal in March.[43] Side by side with these announcements were the next United States domestic system, Hughes Aircraft's Galaxynet, and the regional European system, Eutelsat, launched by Arianespace.[44]

Intelsat had its first major challenge in 1983 from United States companies applying for an exception to the noncompeting-systems restraint of the Intelsat agreement. The Reagan administration had been strongly supporting these applications as part of its overall emphasis on deregulation of telecommunications. This emphasis also manifested itself with the settlement of the Department of Justice–AT&T antitrust suit that same year and the encouragement given to private satellite launch services the next.[45]

April 1984 saw the culmination of a forty-year trend in the launch of GTE's Spacenet 1 by Arianespace: a private telecommunications satellite put into orbit by a nongovernment agency. Coupled with the repair of the Solar Max satellite that spring and the recovery of the two defective satellites in the summer, both by NASA's space shuttle, this breakthrough seemed to herald a new age for commercial satellite information networks.

Instead, the next few years were exceptionally unfortunate for the satellite industry, the technical achievement of the shuttle's recovery flight only temporarily overshadowing the fact of the shuttle's loss of two satellites in January 1984. This twin loss, coming so unexpectedly after years of successful launches, confused the satellite insurance industry and sent prospective users flocking to Arianespace. Optimism there was short-lived as well, with Ariane having its first unsuccessful commercial launch in August 1985, a loss that seemed to initiate a string of other satellite disasters through that September. The January 1984 and third-quarter 1985 losses totaled more than $800 million, at one blow seemingly destroying the satellite insurance industry.[46] At the same, time rapid growth of systems in the previous few years had led to a capacity glut threatening operators' profits as well. The existing international system was strongly challenged by developing countries in the

October 1985 World Adminsitration Radio Conference (Space WARC), adding to reluctance for any new ventures.[47]

Finally, the space shuttle Challenger tragedy in January 1986 put a substantial delay in the ability of the United States to launch any satellites at all, even if these other factors had not been present. Arianespace continued operating substantially behind schedule as well.[48]

The recent shocks underscore the uncertainty always confronting the user of satellite information systems. This book cannot help readers to control such forces, but by appreciating the forces, individuals will be able to make more informed decisions as to the inherent risks and opportunities of these systems.

A BRIEFER TECHNICAL BACKGROUND

This book is not intended to educate the reader in the technical intricacies of satellite engineering,[49] but at least a superficial understanding of the communications principles involved is necessary to appreciate the value and limits of satellite information systems. The elementary discussion includes:

- The basic structure of the communications process,
- Its physical counterparts in a satellite network,
- Communications principles that limit their effectiveness,
- Geophysical principles that limit their effectiveness,
- A general framework for categorizing systems, and
- The importance of satellites for each category.

Structural Elements

Any act of communication can be subdivided into three stages: transmission, when the signal is emitted; processing, as the signal is conveyed to its intended goal; and reception, when the signal finally arrives. No intelligence need be attributed to this sort of communication. It applies just as well to the recital of a poem to an eager audience as to the operation of a typewriter: transmission, as an object presses a particular key on the keyboard; processing, as a series of levers (and probably some electric relays and a daisy wheel) transform the pressure on the key into the extension of a molded character against a ribbon; and reception, as the imprint of the character is transferred to the paper.

Physical Counterparts

Communications over satellite information networks follow the same breakdown: transmission of a signal from the earth to an orbiting satellite; processing of the signal within the satellite, including its redirection back to earth (even if just as a passive reflector of the original signal); and reception of the usually enhanced signal back on earth.

The site transmitting the signal to the satellite (and the signal's progress to that point) is called the *uplink*; the returning signal and the site receiving it are called the *downlink.* Both sites are usually referred to as *earth stations,* a term that also includes those installations used to track and control the satellite itself. A typical earth station employs an antenna system and a solid dish. The parabolic shape of the familiar satellite dish helps concentrate and direct outgoing signals rather than have transmission power wasted in sending out an omnidirectional signal. Similarly, the dish focuses incoming signals into a single point so as to maximize the strength of reception.

Larger dishes thus indicate greater transmission and reception capacity. The more power in the processing component of the communication, that is, in the satellite itself, and the more precise and directable its own equipment, the smaller the dish need be. These processing elements of the satellite are called transponders, which are themselves reception-transmission systems whose most common function is simply to amplify the uplink signal and change its frequency (so as to avoid interference) for the downlink signal.

(Transponders themselves are, of course, complete transmission-processing-reception [though in reverse order] communication systems as described here, just as the downlink earth station, unless it is already on the premises of the end user, itself is just part of a more extended communication loop. The discussions in this book are usually restricted to the earth-to-satellite-to-earth segment of the communication, rather than to a shorter or longer piece of it.)

No matter how precise or directed a satellite's downlink signal may be, the signal is able to be received over a wide geographic area. This area is known as the satellite's *footprint.*

Limiting Communications Principles

Historically, most telecommunications has been carried on by analog transmission; that is, by representation of the signal as a smooth wave approximating the human voice. Computer communications, however, require representing the signal as discrete bits of numerical information; this quantization is called digital transmission. Even for inherently analog signals such as voice and video, digital transmission has advantages in terms of speed, accuracy, and volume, as well as the ability to encode signals for privacy and to package and transmit many different signals simultaneously (multiplexing).[50] Nearly all existing satellite systems employ analog transmission; nearly all proposed ones, digital transmission.

The more accurate a signal is to appear, more information must be transmitted more quickly. For digital transmission, speed can be measured as the number of bits (binary digits) of information conveyed per second. To recreate adequately an ordinary telephone conversation, more than 1,000 bits per second are needed; for high-fidelity music, more than 5,000 bits per second; and for a full-color television signal, more than 1 million. Limits on transmission speed at present are dependent not on the transmission itself but on the speed at which the signal can be reconstructed at the receiving end.[51]

There also is a tension between the ability of a signal to carry large amounts of information—that is, a very high-frequency signal or conversely, one with a very wide bandwidth—and the reliability with which the signal can received. Very high-frequency/wide bandwidth signals, however, are more subject to attenuation in their progress from and to the earth, and to interference from rain and even flying objects. Lower frequencies, which meet less demanding needs, are already being crowded out; newer satellite systems offering more complex services are always pressed to employ higher frequencies, with an attendant need for technical fixes to reception problems.[52]

Limiting Geophysical Principles

The benefits which the geostationary orbit possesses naturally pose potential crowding problems.[53] Typically, the crowding is not physical; that is, there is little danger of actual collision. Geostationary orbit is not a flat band, but a tunnel roughly 1,000 miles wide. In fact, over the continental United States (which is the most physically crowded portion of the orbit), in some cases three satellites will occupy the "same" orbital slot. The extra satellites are in parking orbits and are not broadcasting. They are not broadcasting because satellites interfere with one another (although for a special type of interference, occurring when the ground station, the satellite, and the sun are all in an exactly straight line twice a year, the signal will be briefly rerouted through a parking orbit satellite). They are there at all because satellites do not last forever.

Satellites are placed in geostationary orbit typically 3 degrees of arc apart (with proposed spacing of 2 degrees[54]) to avoid confusion of their signals. Two satellites transmitting side by side at the same frequency would produce a hopeless jumble on the reception end. Nonetheless, particular locations and particular frequencies are obviously attractive. Open competition for slots has largely been avoided because technological advances have leapfrogged telecommunications demand and because satellites ultimately leave geostationary orbit.

In an idealized physics textbook illustration the earth would be a perfect sphere, there would be no outside gravitational forces, and satellites in geostationary orbit would remain there for all time. The earth is not a perfect sphere, however, (in fact, it has a pronounced bulge about the equator); the gravitational fields of the sun and moon, although small compared to the earth's, are not negligible; and from their first day in orbit geostationary satellites may waver.

Small corrections, called station keeping, are constantly needed for the satellite. Similar corrections are necessary for the satellite's attitude toward the earth so that antennas are always facing in the right direction. More complicated attitude control, pointing the satellite's solar batteries toward the sun, is also required. These batteries power the satellite's communications tasks (aided by storage batteries when the satellite is in the earth's shadow), but they are inadequate for station keeping and attitude control. Fuel carried as part of the satellite's payload must be used for these functions, and when it is exhausted the satellite loses its geostationary advantages long before it is in any danger of actually falling out of orbit. The average life of a geostationary satellite is roughly seven years (with a shorter span for all other types of orbits), although some have lasted nearly twice that long.[55] Future plans for the industry often concentrate on refueling possibilities or redirecting reception dishes as means of extending useful satellite life indefinitely.

A General Typology

Throughout the remainder of the book, all satellite information systems are divided into three different types with quite distinct economic, technical, and legal attributes.

1. One-to-one information transfer, going from a single transmission to a single reception, as in a typical telephone conversation;

2. One-to-many information transfer, going from a single transmission to multiple receptions of the same signal,

often without a formal downlink station, as in a typical radio broadcast; and

3. Many-to-one information transfer, going from multiple
    transmissions to a single reception of many signals, almost
    always without a formal uplink station, as for a weather
    satellite.

Criticality of Satellite Information Systems

For each type of information transfer, the necessity of a satellite information system differs; it is odd that the need is in inverse proportion to
the value of the market served. Satellite information systems are a cost-
effective means of providing one-to-one information transfer, but they
have many successful competitors in what is the largest part of the information transfer market. Satellite systems are the dominant mode in
the somewhat smaller one-to-many information transfer markets and
will likely increase their participation. For many-to-one information
transfer, satellite information systems are the only available means, but
at present represent negligible revenues unless their inestimable military
value is included.

Chapter 2
One-to-One Information Transfer

DEFINITION

The first major use of satellite information systems is one-to-one information transfer; that is, the transmission of a single signal to a single receiver. The most familiar example of one-to-one information transfer is a telephone call.

One-to-one information transfer can be differentiated from one-to-many (chap. 3) and many-to-one (chap. 4) in at least four ways. First, the uniqueness of the signal in one-to-one transfer constrains the system and limits any possible economies of scale. A television broadcast (one-to-many) can reach a larger audience by using a more powerful transmitter; a remote sensing satellite (many-to-one) can cover a larger area by sacrificing some fineness of detail. A telephone system cannot connect more people by a simple increase in power, and voice communications (much less data communications) can afford little degradation or mixture without losing their value altogether.

Second, one-to-one information transfer is typically two-way, or at least has the capacity to be so. Every communication "closes the loop" between sender and receiver and, in fact, eliminates any practical difference between sender and receiver from moment to moment. This symmetry of relationship places additional burdens on the system.

Third, if every sender is a receiver and vice versa, this means that every uplink is also a downlink. A television broadcast has most of its power and processing on the uplink side, with no physical connection needed to the viewer's home; conversely, the wheat crop takes no special action to be visible to the remote satellite that, in turn, has a complex downlink segment. For conventional telephony, some physical connection must be made to each interchangeable sender-receiver.

Fourth, the total number of individual broadcasts in the world is on the order of 100; all remote sensing satellite systems collectively, including the military, probably are not capable of taking as many as 10,000 images daily. The number of communications over telephone systems is more on the order of a billion a day. While only a small portion of these involve satellites, this level of volume still means that one-to-one information transfer is the dominant use of satellite systems.

HISTORICAL AND TECHNICAL BACKGROUND

Technical Limits

One-to-one information transfer (telephony) has modest technical bounds on performance compared to one-to-many (broadcasting) and many-to-one (earth observation) applications; nonetheless, satellite systems strain these limits.

The bandwidth required to capture the human voice is only about 3,000 Hz (compared to about 4 million Hz for a reasonable video signal).[1] Orbiting satellite systems are capable of handling roughly 30,000 voice channels simultaneously, and the next generation will have an 80,000-channel capacity.[2] This volume of traffic is needed to support the economics of satellite communications; it also largely eliminates the advantages of narrow bandwidth and increases the possibilities for "cross-talk" (interference across channels).

Similarly, voice communications (although not data) require a relatively low level of reproduction quality; high-fidelity music is not expected from the telephone. While atmospheric interference during the uplinks and downlinks to a satellite system is probably less (thanks again to the narrow bandwidth) than that caused by repropagation of the signal on the ground through microwave relays, what interference does exist is magnified by the power amplification needs of satellite systems.

Interactive voice communications do not need to be particularly rapid. Even at the speed of light, however, a signal takes a half-second to reach and return from a satellite in geosynchronous orbit. Although some users are very sensitive to this built-in lag in response time, the vast majority seem not to notice it at all. In any event, the AT&T network for domestic communications, traffic permitting, typically pairs a satellite circuit in one direction with a ground circuit for the return; with this configuration, the lag is an almost imperceptible quarter-second.[3]

As satellite systems follow the trend of ground systems toward conversion to digital (rather than analog) processing, the consequences of all these limits will be ameliorated. Digitization can address the volume problem through multiplexing, and accuracy difficulties by both requiring less power and incorporating error-checking codes; while digitization cannot change the laws of physics and speed up interactive communications by satellite, it can increase the rate of one-way communications dramatically.[4]

Historical Development

Although the United States is virtually the only country that has a private corporation run its telephone system, that corporation, AT&T, had until recently an almost total monopoly on long-distance service, and the company has been closely regulated throughout its life. Advances in technology have periodically burst the apparent bounds of regulation, only to be reined in quickly.

As national provider of telecommunications services, AT&T carried on much of the communications research responsible for these advances. In the early days of the company's history the consequence of improved telephone service led to an open war with the telegraph system, a war AT&T won.[5] Long before radio (as early as 1893), telephones were used to transmit music and news bulletins in Hungary[6]; early AT&T leads in radio technology in the United States were bartered away, under government pressure, to other firms.[7] The same was true for motion pictures[8] and television broadcasting.[9] In each case, antitrust and other competitive considerations led to further implementation of the technology by other companies—in fact by whole new industries. When the same point in the development of satellite systems was reached, however, the government itself took over.

Not only did AT&T develop the technology for communications satellite systems[10]; as part of national defense efforts in the 1950s it was heavily involved in developing launch vehicles as well as in designing and implementing the communications and tracking systems used by the military in operating the earliest satellites.[11] On this basis, AT&T asked NASA in December 1959 to be given effective control of satellite communications with the exclusive right to build and operate networks at its own expense, subject only to continued NASA and FCC regulation.[12]

Although this request was turned down, the next administration authorized a joint venture of AT&T with NASA to develop and launch Telstar, the first active communications satellite. After the successful launch in 1962, AT&T attempted to argue that communications satellites were not part of the national space program, but just "a microwave system in the sky"; instead, within weeks, the quasi-governmental corporation ComSat[13] was formed (with AT&T holding a major share) to carry on satellite network development.

ComSat's mixed nature (the AT&T chairman called its enabling act "about the most bobtailed piece of legislation I ever saw"[14]) was only modestly successful in creating a domestic network, and in 1972 the FCC allowed the creation of independent private networks. The AT&T satellites were precluded by this decision from offering any services other than public long-distance services until three years after first launch.[15] Immediate responses to the FCC ruling were satellite systems carrying cable television signals. The first two AT&T satellites were launched as a joint venture with GTE in 1976, with delays due to both the regulatory restriction and the still prohibitive costs of satellite transmissions and earth station operations. Although this first pair of satellites was very much underused, a third was launched in 1979 just before the regulatory limit expired, to take advantage of the anticipated windfall to follow.[16] This optimism was vindicated by the opportunity to provide private line and other dedicated services as well as broadcasting signals for the national television networks, all accompanied by radically decreased ground station costs.

In the international arena, the progress of Intelsat in providing one-to-one telecommunications services was much smoother and more successful until quite recently.[17]

ALTERNATIVE USES OF SATELLITE SYSTEMS

One-to-one information systems can be characterized by the type of information transmitted: voice, data, or video. In 1985 an estimated breakdown of one-to-one satellite traffic by these categories was voice (74%), data (18%), and video (8%).

Voice

One-to-one voice communications, sometimes denigrated as "plain old telephone service" (POTS), still presents some special opportunities that can be augmented by satellite systems. The voice equivalent on an on-line data base is the provision of Muzak signals over telephone lines.[18] Ordinary voice signals are also used as part of passive alarm systems, where a telephone call activates an alarm in a remote location.[19]

More significant than these tangential applications is the audio-conference. In its least sophisticated form, a conference call is set up among several users simultaneously; in areas served by electronic switching, users can initiate such arrangements without any operator intervention. More formal arrangements, with microphones at each location (to control unwanted transmissions) and where the communications proceed in a more structured way, have been especially successful in carrying on "tele-education": use of satellite systems to bring instruction, typically at the college level, to a region or even an entire nation at the same time.[20] In developed countries, audio-conferences are used much more for corporate training courses than for public education, although outstanding examples exist such as the University of Quebec system (linking ten sites) and the University of Wisconsin Educational Telephone Network (with more than 200 locations).[21]

In audio-conferences for business rather than educational purposes, the identity of the speaker becomes important. One method for dealing with this problem is the remote meeting table (RMT) system, where participants at one location are put in position alternating with "phantom" places, occupied by a microphone, that correspond to the position of parties at the other location.[22] Each speaker then is automatically identified by his "place" in the room. This arrangement provides a very effective simulation of a face-to-face meeting without the expense associated with a full-scale video-conference.

Data

One-to-one transfer of data is almost always a one-way process. The most common applications, in increasing order of importance, are as an adjunct to audio signals (e.g., telewriting), a substitute for video signals (e.g., facsimile), or as a mode of man-machine or machine-machine communications.

Telewriting is as old as the telephone itself, and is a simple marriage of the telephone and telegraph.[23] A stylus is connected to a two-dimensional harness, with its movements reduced to an electronic signal and communicated to a parallel hookup at the other end.

A more sophisticated version of remote writing, developed nearly a century later, is the electronic blackboard, created by AT&T and first used at the University of Illinois in 1974.[24] In this application, writing on what appears to be a normal blackboard is conveyed over normal telephone lines to a video monitor at the receiving location.

These types of transfers are useful as a supplement to audio transmissions for business and educational purposes, but are not appropriate for conveying data alone, such as a document or photograph. For this, facsimile transmission is more effective. In this process, an optical scanner converts each position on a page to a binary choice of black or white, with this series of numerical information then transmitted to the end user to re-create an individual picture. This process is highly efficient if compared to actually retyping a document, and can be carried on without human intervention. It takes anywhere from 5 to 30 seconds to transmit a single page, with gains in speed associated with transmitting only the locations where the signal changes (i.e., most of a page is white).[25] With the decline in reliability of postal services, coupled with a premium on speed, the facsimile service has probably been the fastest growing satellite application of the past few years. Satellite Business Systems (SBS) launched a digital satellite network beginning in 1980 that can provide image transmission at the rate of nearly one page per second; Xerox's XTEN system floundered on problems associated with its ground distribution system, but was on the verge of enormous success allowing its own reproduction machines to be linked to a satellite system.[26] Federal Express, operating on transponders leased from GTE Spacenet satellites, virtually invented a new industry with its Zap Mail facsimile service; the anticipated volume did not occur, however, and the service was stopped in 1986.[27]

Data transmission from computer to computer or computer to end user is the largest (with respect to volume) use of satellite networks in one-to-one information services. The latest digital systems can send data at a rate of 3.5 million bits per second, 1,000 times faster than conventional voice lines.[28] With this speed and the absence of a need for real-time interactive communications, satellite data transfers are also growing significantly faster than voice transmissions.

Video

The first actual videophone was made long before it became a staple of science fiction, when AT&T exhibited a working model at the 1939 World's Fair.[29] Although a videophone-like device has achieved some popularity as an intercom/security monitor, various attempts to offer videophone service on a national basis have always fallen short due to the costs of the required video camera and enormous bandwidth. AT&T attempted to market a Picturephone system to business in the late 1960s, but was unable to achieve sufficient market penetration to be worthwhile.[30]

Instead, Picturephone survived in a new identity as the basis of a video-conference service. AT&T's Picturephone meeting service, first offered in 1974 (but anticipated by a similar British system, Confravision, in 1970), provided Picturephone-equipped studies in major cities for use by private clients.[31] This effort was not tremendously successful for the company itself, but did demonstrate the value of video-conferences to its business customers, enough so that today nearly all the major hotel chains (Holiday Inn, Hilton, Intercontinental, Marriott, Sheraton, and Hyatt) offer national video-conference facilities relying almost exclusively on satellite systems, joined in similar systems by a number of large corporations (e.g., Ford, General Motors, IBM, Aetna, Du Pont, Merrill Lynch, and, of course, AT&T itself).[32]

Adding a video signal to two-way voice communications has been wildly successful. A meld in the opposite direction—adding interactive responses to a video signal—has failed to receive as wide implementation. Such services, generically called videotext (as opposed to one-way services collectively labeled teletext), in effect use the telephone system to connect enhanced television sets with computers storing information that can be displayed, on user request, on the screen. Experimental systems, usually involving "shop at home" systems, have been implemented in Britain, France, Canada and the United States without dramatic user response.[33] The most famous, or notorious, system, however, was a viewer-polling network allied with a cable television system, named QUBE; in this application, the pervasive regulatory and economic concerns, low levels of viewer participation, and failure to lure subscribers to other services limited its expansion.[34]

THE AT&T DIVESTITURE

The Road to Divestiture

The regulation of AT&T's monopoly over telephone services, as discussed earlier in this chapter, had involved precluding the company from applications of communications technology in related fields: radio, motion pictures, television, and ultimately, communications satellites themselves. In the 1960s and 1970s, attempts to segregate computing from communications led to a series of hairsplitting decisions by the FCC that ultimately ended in failure.[35] In 1966 the FCC initiated its first Computer Inquiry study (CI-I), intended to evaluate the impact of advances in the computer industry on the AT&T telecommunications network, even though Western Electric, the manufacturing arm of the Bell system, had been debarred by a 1956 consent decree settling an earlier Department of Justice antitrust suit from producing any computing equipment. This study was prompted by the recognition that the electronic switching systems (ESS) for managing the telecommunications network, which were first implemented by the Bell system at that time, were computers in all but name. It was feared that the power of these systems, coupled with an officially sanctioned monopoly over telecommunications services, would result in an unacceptably strong stranglehold over the entire industry. Although CI-I ended in 1972 with a determination that no additional regulation was required, the intervening years saw many regulatory decisions intended to increase competition for all offerings except local, unenhanced services.

Reaction to the piecemeal regulatory approach of the FCC—protect the monopoly, divest the technology—took both judicial (a Department of Justice antitrust suit in 1974, the third major suit in the company's history) and legislative (the first version of the Consumer Communications Reform Act in 1976) forms: divest the monopoly, and no one will care about protecting the technology. Not to be outdone, the FCC initiated a second Computer Inquiry (CI-II) in 1976. The upshot of all these initiatives was:

- In 1980 CI-II was completed with a determination that basic services, those not benefiting from any advancing computing technology, including ESS capabilities, should remain regulated, with AT&T having an effective monopoly, while enhanced services (receiving such benefits) should be open to competition.
- In 1982 the Department of Justice antitrust suit was resolved with a decision to divest AT&T of any interest in its local operating companies (the statewide or regional entities providing local service and connection to the national and international long-distance networks), with the option to manufacture telecommunications equipment, after-last minute negotiations and adjustments, remaining with AT&T.
- In 1982, given the above results, all legislative efforts were abandoned.

Bypass and Satellites

What do issues of antitrust philosophy and political motivations behind the AT&T divestiture have to do with satellite systems? The clue is in the claim, still contested after divestiture, that AT&T long-distance rates did—and should—support the expenses of investment in the local network that makes all communications possible. When the United States telecommunications network was a seamless web, paper transactions between AT&T and its wholly owned local subsidiaries were not of fatal importance. After divestiture, a fixed charge was imposed on all long-distance suppliers, including AT&T's competitors, for connection with the local network.

Consider, then, the positions of the major parties to this melee:

- AT&T is confronted with making real-dollar payments to often hostile former subsidiaries that, in turn, have control over access to AT&T services.
- AT&T competitors are given unparalleled opportunities to compete with AT&T, together with unprecedented payments to the local companies.
- The local companies, legally precluded from indulging in either long-distance services or equipment manufacture, are dependent upon these access charges for their economic survival.

What weapon does each of these potential antagonists/allies have against the other? Satellite systems can avoid the necessity (or detectability) of the connection of the long-distance and local systems. This direct interconnection of an end user with long-distance service is called bypass of the local exchange.[36]

Bypass technology is usually represented by rooftop satellite dishes operated by large businesses or governmental agencies effectively to create their own private networks.[37] Over and above the direct incentive of avoiding access charges, such users may be motivated to take advantage of the digital services and the greater speed afforded by new satellite systems.[38] Nonetheless, the growth of bypass places additional pressure on local rates due to the lost access charges, which, in turn, creates additional pressure for bypass in a vicious cycle. The issue is currently under intensive FCC study.

OTHER PRIVATE SYSTEMS

Mobile Phone

A mobile phone requires no physical connection to a network and, so, can be used while traveling. Until recently, the only means of providing such service was cellular radio: dividing up a given receiving district into cells, each of which is served by an individual radio tower for picking up and transmitting signals, with these towers then joined to some long-distance network.[39]

A clear alternative to cellular radio, with its low number of customers that can be handled, its geographic limitations, and its need constantly to "pass off" signals from one cell to the next, is a satellite system. After divestiture, twelve companies applied to the FCC for permission to create such a satellite system.[40] An application available on many of these proposals, which has already aroused considerable interest, is that of a national paging system. In Canada, a national mobile satellite system has already been approved, with anticipated deployment in 1990.[41]

In one arena mobile satellite service already exists. In 1982, the International Maritime Satellite Organization (Inmarsat) took over and radically expanded a satellite system for one-to-one information transfer by ships at sea.[42] Currently under development are plans to extend Inmarsat's telephony services to planes in the air as well.[43]

The Teleport

A rooftop antenna for local bypass is a relatively modest form of private network. Much more ambitious is the notion of creating an entire business complex dedicated to capitalizing on the new satellite technologies. The prototype effort of this sort is Teleport, a satellite communications center on Staten Island in New York City, developed jointly by the Port Authority and Merrill Lynch.[44]

The Teleport is connected to the rest of New York City and to Hoboken, Jersey City, and Newark, in New Jersey, by a fiberoptic cable network. At the Teleport itself, ultimately seventeen earth stations will be set up (three are in operation now) to provide connection with long-distance telephone services, high-speed data links, video-conferences, and facsimile reproduction, all through satellite information systems. The New York Teleport is seen as the forerunner of similar installations in other major cities.

ALTERNATIVES TO SATELLITE TELEPHONY

Conventional, twisted-wire pair technology is not an effective competitor with satellite telephony on the basis of capacity alone. Such competitors do exist, however, in fiberoptics for national systems, and transoceanic cables (especially when employing fiberoptics) for international systems.

Fiberoptics

Fiberoptics refers to the new technology of carrying information in a beam of light traveling along extremely thin glass fibers. This technology has the ability to carry thousands of times more information (relative to the size of the cable) than traditional electrical signals on copper wires. Fiberoptic capacity is so great that the bandwidth exists to carry as many video signals as older systems can carry voice signals.

If a fiberoptic network were operating at full capacity, it would almost certainly be cheaper to send signals by that means instead of by satellite. The difficulty lies in reaching full capacity. Fiberoptic systems are extremely expensive to build and, unlike satellites, are distance sensitive: each additional mile the signal must travel requires an additional mile of cable. Therefore, fiberoptic systems at the present only make sense for a few relatively short but high-volume routes, such as a planned New York–Washington system. At the same time, the construction costs of, and investment required for the right-of-way for, fiberoptics are highest in the dense urban centers where they are most justified.

Industry opinion remains sharply divided as to the degree that fiberoptics will ultimately supplant satellite systems.[45] While it is clear that satellites will remain the medium of choice for low-traffic and/or longer-haul routes, the definitions of "low" and "long" are quite uncertain. A common rule of thumb is cited by Federal Express's satellite manager: if it is more than 400 miles away, go by satellite.[46] At the same time, plans are being drawn up for a 700-mile New York–Chicago network.

Satellite systems also have the strong advantage of existing here and now. Every year that passes before implementation of fiberoptic alternatives makes the conversion that much more expensive and hence more unlikely as well. The reality is that a strict satellite *or* fiberoptic decision is a false dichotomy. The few present (and most proposed) systems are usually hybrid networks attempting to integrate satellites and fiberoptics to capitalize on the advantages of both.[47] A good example is the Teleport network discussed above, where a very short but high-traffic route (Manhattan to Staten Island) is handled by fiberoptics and then connected with a number of different satellite systems for long-haul traffic and specialized services.

Transoceanic Cable

The most direct threat to existing satellite systems is the embedding of fiberoptics in the next generation of transatlantic cable (the TAT-8), intended to go into operation later this decade.

The older transatlantic cables providing telecommunications services between Europe and the United States are less cost efficient than the current generation of satellite systems. Nonetheless, for some time ocean cable has been seen as a valuable alternative providing backup reliability and insurance; there is also a lingering affection for such systems because the cables are owned outright and managed by the cable operators, as opposed to Intelsat's administration of satellite systems.[48]

TAT-8 will reverse the economic advantage in cable's favor, with satellites' superiority confined to speed of data transmission and some flexibility in responding to changes in traffic volume.[49] This advantage will presumably endure until a new generation of satellite systems technologically leapfrogs cable once again. Perhaps an even greater threat exists in the creation of the first transoceanic cable across the Pacific, under discussion for 1989 operation, where satellites face no competition at present.[50]

TYPICAL APPLICATIONS

The range of typical applications of satellite information systems in one-to-one information transfer can be suggested by an overview of some approximate costs followed by five brief case studies.

Costs

Costs of providing ordinary voice communications by satellite are almost impossible to estimate directly since an end user has no choice as to whether his signal proceeds over satellite or not. The charges to him are based on an average cost over the entire network, terrestrial, and satellite portions weighted together. A range of relevant costs can be appreciated from the fact that Intelsat recently sold nine transponders for $16 million, that is, under $2 million each, a price thought incredibly low; the next proposed auction expects to charge $3 to $5 million per transponder.[51] (In rough terms, each transponder is capable of handling 1,000 voice channels at once.) Lease of a transponder is roughly $1 million per year or $100 thousand per month, with wide variation in prices depending upon the preemptability of the signal.[52]

Facsimile

Federal Express's Zap Mail facsimile service was formed after it was recognized that nearly half of its overnight deliveries consisted of documents that could be transmitted electronically.[53] Starting in 1985, Federal began offering facsimile transmission over high-speed data lines leased from AT&T, slowly switching to transponders leased on GTE's Spacenet satellites.[54] Federal planned to use its own satellite, Expresstar, for a system that, when combined with Zap Mailer receivers ($5,000–$10,000) at the end user's location, will deliver facsimile reproductions at about twenty-five cents a page; without such receivers, Federal's charge is a little over a dollar a page.[55] Federal also planned to offer international Zap Mail service by piggybacking on Intelsat circuits late in 1986.[56] Federal nonetheless dropped the service in 1986 for economic reasons.

Video-Conferences: The Corporate Route

In 1981 Hewlett-Packard (HP) committed to construction a corporate teleconferencing network to handle communications with its 79,000 employees in seventy countries.[57] A monthly video-conference in 1982, transmitted by satellite to provide sales training nationally, quickly grew to a several-times-a-week system over sixty-five receiver locations addressing customer training, internal problem solving, corporate communications, administrative meetings, and formal employee education.[58] Savings in travel time alone were estimated to pay for a dedicated network in one year. More generally, transmission costs are about $1,000 per hour for some 210 corporate video-conferences systems existing in 1983.[59]

Video-Conferences: The Political Route

The latest beneficiaries of advances in video-conferences have been politicians. In 1978 the Alaska legislature authorized the construction of a statewide tele-conference network linking remote areas within the state with each other and with Washington, D.C.[60] Campaigning is not allowed on the network, which is used to inform residents of actions in the national and state capitols and to provide immediate feedback to representatives.

Use of video-conferences became an essential part of national campaigns in 1984, allowing candidates to "appear" at several locations simultaneously, and for centralized presentations by the major parties to be made to the entire country.[61]

Bypass

The economic advantage and regulatory uncertainty associated with local bypass is so great that even AT&T itself has entered a joint venture with British Telecom to provide rooftop-to-rooftop communications between New York and London.[62] Avoiding the local loop is expected to cut the cost of a transatlantic call in half, if customer traffic is heavy enough to support the multimillion-dollar investment required for a private earth station network.

SBS: THE INTEGRATED SATELLITE NETWORK

The wave of the future is probably best represented by the Satellite Business Systems (SBS) network, which launched its first satellite in 1980. The SBS program of services is as follows.[63] A single communication system is provided for transmitting voice, data, facsimile, and video signals, all in a completely digital network at high rates of speed. A ground user's earth station, costing about $250,000, would be queried fifty times a second for requests, and the whole system would operate as a giant time-sharing computer. The SBS network is the marriage of computing and communications in satellite networks that has been artificially delayed for almost 20 years.

Chapter 3
One-to-Many Information Transfer

DEFINITION

The second major use of satellite information systems is one-to-many information transfer; that is, the transmission of a single signal to many receivers simultaneously. The most familiar examples of one-to-many information transfer are radio and television broadcasting.

There are two enormous practical differences between one-to-one and one-to-many information transfer. First, the former (telephony) is essentially two-way communication, while the latter (broadcasting) is unidirectional. Second, the number of different signals propagated at any one time by one-to-one networks is on the order of one hundred million times greater than for one-to-many transfer, in part due to technical demands on video transmissions. These differences taken together—one-way communication of a small number of signals—mean that distribution and reception of one-to-many transmission is relatively straightforward. No formal downlink station may be required at all, and the system aspects of the communication often are transparent to the user.

HISTORICAL AND TECHNICAL BACKGROUND

Technical Limits

Transmission of video signals has three technical constraints that, until quite recently, limited the number of channels theoretically possible. First, representation of an adequate video signal requires very large bandwidth of about 6 million Hz (cycles per second), with a single television channel using up more of the frequency spectrum than 2,000 telephone calls. At the same time, this large bandwidth also means that the signals are especially prone to interference from nearby (in frequency terms) transmissions, so that even more space must be reserved. Estimates of the number of channels allowed by this limit have fluctuated with both technological and political shifts. The first regulation of channel frequencies by the FCC allocated thirteen channels for broadcasting; over the years, this number has grown to twelve, then eighty-two, and then sixty-eight.[1] This technical limit has never become a practical one due to the limitations of programming availability. Second, these high frequencies require a great deal of power. Transmission over wires requires amplification of the signal every few hundred feet (as is the case for the final segment of cable television delivery).[2]

The solution to this problem is to place the video signal on an even higher-frequency radio carrier signal which, in turn, creates a third set of difficulties. The audio part of the signal is controlled by frequency modulation (i.e., FM radio), and is less subject to interference than the amplitude-modulated (AM radio) video signal.[3] Further complicating matters is the inability of the video signal, even at high frequencies, to convey enough information to update the entire picture instantaneously. In the United States, early technical limits had half the image re-created every one-twentieth of a second. Later European systems began with accommodating more rapid updating, with the result that both broadcasts (and television sets) developed for one set of standards are incompatible with one another.[4]

Historical Development

The first radio networks were formed in the 1920s by AT&T, CBS, and RCA (the parent corporation of NBC, the National Broadcasting Company), although both of AT&T's competitors used its lines to transmit their signals. Litigation with RCA and the Federal Communications Act of 1934 forced AT&T to forego direct involvement in radio broadcasting while retaining its distribution role. The first applications for the thirteen television channels authorized by the infant FCC were snapped up by the two remaining radio giants, who, not incidentally, owned most of the proprietary rights to video technology. The first regularly scheduled commercial broadcasts did not begin until 1939 and were quickly interrupted by World War II. At the end of the war, CBS, RCA, and the new American Broadcasting Company (ABC) pursued television broadcasting aggressively. ABC had been formed in 1943 from a portion of NBC by an FCC-Justice Department divestiture order; from this disadvantaged start, ABC has run a weak third among the networks for almost forty years.

The capital resources of the three national networks and a freeze on new licenses allowed them to dominate the television industry. When the number of authorized slots increased from twelve to eighty-three in 1952, the existing first generation of sets was unable to receive the new UHF (ultra-high-frequency) stations. Accordingly, these channels and their programming never had an opportunity to develop a following; in 1962, when the FCC required all new receivers to be capable of receiving these signals it was already too late, and today the scant programming offered on them is almost entirely noncommercial.[5]

Also in the early 1950s, AT&T began to operate its long-distance communications on a terrestrial microwave network, and the television networks followed suit,[6] beaming their programming nationally over the telephone network, then rebroadcasting locally at higher frequencies over the air directly into homes within line of sight of the rebroadcast antenna.

Although some of the very first satellite systems were used to broadcast special events, it was not until 1972 that RCA made the first regularly scheduled use of satellite transmissions, oddly enough by leasing transponder capacity on the Canadian satellite broadcasting system, ANIK-A, which had become operational the year before.[7] American domestic systems arose in the late 1970s, but were typically used for special events and news programming only. Once again the television networks waited for AT&T's system to become operational in 1980, and it was not until 1982 that the bulk of network programming was routinely carried by satellite.

Why Broadcast by Satellite?

Were someone to design a broadcasting network from scratch today, transmission by satellite would be the obvious choice for a long list of reasons, including insensitivity to distance, increased capacity, high quality of signal, and flexibility of programming.[8]

The biggest advantage satellite broadcasting networks possess is their insensitivity to distance. Terrestrial microwave propagation requires repeating stations every few miles as the necessary line-of-sight signal is lost beyond the curvature of the earth; a single satellite signal can reach the entire continental United States simultaneously for only a fraction of the ground network cost.

This same insensitivity to distance (more precisely, the absence of involvement with any physical system of amplifiers and relays) also leads to the enhanced capacity and signal quality attributes of satellite systems. The number of relays needed places an intrinsic limit as well as high costs on the number of different signals that can be carried, and each additional amplification and repropagation of a signal inevitably brings about additional degradation. Satellite systems with a single transmission required for each signal are free from both these constraints.

In turn, wide geographical coverage, large capacity, and high signal quality make it easier to develop hybrid networks and to program more flexibly, given this ability to send many signals accurately almost anywhere. A satellite system operator is at the same time theoretically free to choose among programming originating almost anywhere for retransmission.

Given these enormous advantages, why did it take so long for the television networks to embrace satellite systems? Part of the delay was certainly due to the long regulatory road to authorization of commercial satellite systems, and a larger part to the networks' longstanding familiarity with and immense investment in AT&T's terrestrial system. Despite the advantages of satellite systems, however, the three national networks still probably would not have converted if not for the last decade's growing competition from alternatives to network broadcasting, especially cable television.

CABLE TELEVISION: AN ALTERNATIVE TO NETWORK BROADCASTING

Localism and the Domination of the National Networks

Explicit FCC control of television broadcasting is based on the concept of localism, that is, that the public interest is best served by local community control and production of broadcasts. This notion is partially due to traditional theories of democracy with idealized and unrealized hopes of town meetings and church events being televised by the local entity. Nominally, every television station is required to report to the FCC annually on the needs of its local community of viewers and on the station's response to those needs.[9]

A stronger motivation behind localism was the desire, also unrealized in the end, to keep the dominant radio networks from becoming the dominant television networks. No entity is allowed to own more than five local stations. However, only about 100 of the roughly 725 commercial stations broadcasting today are not affiliated with one of the three national networks; that is, routinely offering programming not provided by the networks.[10]

How did the networks' domination arise despite the regulatory bias against it? The answer lies in the resources they possess (economies of scale) and the number of different operations they control (economies of scope). Development of programming is an extremely expensive enterprise that is supported by advertising revenues, as well as later syndication of the programming for rebroadcast. Only a national advertising base can support most prime-time budgets. When these scale economies were coupled with the embedded base of transmission and broadcasting facilities already controlled by the radio networks, as well as the communications and reporting systems in place for coverage of news and sports events, a mere local station had no real opportunity to compete at all.

The networks also quickly dominated the local stations by the variety of different functions they performed. First, they had almost all the capacity for broadcasting live events, the transmission of which even the FCC saw as a special function of theirs. Second, the networks were almost the exclusive providers of scheduled programming for the local stations. Third, they were the effective broker between advertisers and the local stations, with the networks guaranteeing to purchase large blocks of local time at bargain rates for resale to national accounts.[11]

Finally, the last nail in localism's coffin is the fact that the last television broadcasting license issued by the FCC was in *1952*, at which time the FCC decided that all possible channels were permanently allocated. Thus the early initiatives of the networks were solidified and protected from any direct encroachment by future technological advances.

For viewers, this also meant that any area not within line of sight of an antenna authorized in 1952 would be unable to receive any broadcasts, at least until cable television and satellite dishes became available.

The Origins of Cable Television

Cable television is now the name loosely given to community antenna television (CATV). CATV was a response to the unavailability (or poor quality, due to "ghost" signals reflected by interfering objects) of television broadcasting in specific localities. A CATV developer would construct a large receiving antenna (the community antenna) on the highest point in the area, and then connect each household directly to the broadcasts received by a coaxial cable. Fees for connection covered antenna costs.

Transmissions by coaxial cable have significant advantages and disadvantages relative to over-the-air broadcasts. Laying out a cable system is very expensive. It is also financially risky, since it is necessary to create the majority of the network at one time if costs are to be controlled at all, but this must be done without knowing the number of actual subscribers. Television signals also require frequent boosting when traveling across cable and must be amplified every few hundred feet.

On the positive side, in addition to the near-perfect reception afforded viewers is the broadcast capacity of cable. There is essentially no theoretical limit to the number of different signals that may be carried simultaneously on coaxial cable. Thickness of the cable imposes some practical limits, but even the first-generation systems were capable of carrying ten signals, many more than were available; today's systems can deliver more than sixty channels.[12]

Cable systems are potentially two-way systems over which interactive services may be offered.[13] A returning signal requires the same amplification as an incoming one, however, and until recently the periodic amplifiers on most cable systems were capable of working in only one direction. Current systems allow the possibility of two-way amplification or, more rarely, lay a second return cable simultaneously with the first incoming connection.

In any event, the first CATV systems were experiments in the late 1940s, designed solely to improve reception. Several locations now claim credit for being the first to be wired but, by the mid 1950s, several thousand local systems had been created, largely in response to the freeze on broadcast licenses. These systems limped along for twenty years on the edge of solvency. In the late 1960s and early 1970s, several experiments in pay television were conducted in which customers were charged either a per-view or blanket monthly charge to receive recent movies and live sports events. These opportunities were intended to enhance the value of local cable systems, which were usually in broadcast-poor areas to begin with. These experiments typically lost money, being unable to drum up a large number of subscriptions, while at the same time being forced to deal separately with hundreds of fragmented systems.[14]

Cable Discovers the Satellite

While United States broadcasting remained linked to 1950s technology and a 1930s network configuration, Canada's broadcast system, Telsat, had the comparative advantage of starting from scratch. Given its very sparsely settled and relatively rugged terrain, CATV systems were the only way of delivering broadcasts to much of the country. At the same time, the disadvantages of operating small, independent systems were far greater than in the United States. The Canadian government solved both problems by providing the first regularly scheduled national network (Canadian Broadcast Corporation) broadcasts through its ANIK system of satellites beginning in 1972. A central transmission point sent signals to the satellite, which in turn rebroadcast them to all of Canada simultaneously: no reflected signals, and everyone within line-of-sight with the sky. These signals were picked up by a local television receive-only (TVRO) earth station antenna, and sent out over local cable networks.[15] The first United States use of satellites for occasional broadcasting was actually over transponders leased from the Canadian system.

In the early 1970s, Home Box Office (HBO), a subsidiary of Time, Inc., was organized as a pay television service to provide recent movies to Time's chain of cable systems, American Television and Communications (ATC). The service limped along, generating 100,000 customers but still remaining unprofitable. Inspired by the Canadian example HBO decided in 1975 to transmit its programming over satellite, rather than provide videotapes to each location or employ expensive land lines. At the same time, the FCC eliminated regulation of small TVRO earth stations (as opposed to rebroadcasters) and HBO undertook to assist cable operators with their earth stations if they would carry the HBO service.

HBO quickly acquired a million customers and became profitable almost overnight through its decision to broadcast over RCA's SATCOM I.[16] This discovery of the benefits of satellite broadcasting had at least three effects. First, cable systems could now in theory carry any signal, not just those locally available. Second, competition for attractive programming transformed the industry in just a few years so that programming entities effectively controlled the local operators, rather than the other way round. Third, the cost benefits of satellite transmission were so great that almost all cable networks use satellite systems for their delivery. But just as the new technology undermined the domination of the national networks, it contained the seeds of cable's own destruction.

DEATH OF THE DOWNLINK: OTHER FORMS OF SATELLITE BROADCASTING

CATV systems circumvented the national networks' former distribution technology by use of satellites to carry a signal to a TVRO earth station. In turn, satellite transmission makes it easier to circumvent the cable operation by making a formal downlink less and less necessary. In increasing order of freedom, some other forms of satellite broadcasting are master antenna television, low-power television, private satellite dishes, and direct broadcasting satellites.

Master Antenna Television

Master antenna television (MATV) systems are small private CATV systems. They generally consist of some closed environment (an apartment building, housing complex, or hotel, usually; less commonly a restaurant, bar, or library) of a single sophisticated antenna arrangement, most often a TVRO station, with which to deliver and control broadcasting to all units in the complex.

Unlike cable systems, there is usually no direct fee to the user, with the major exception of pay television in hotels. Instead, the system's costs are built into the construction and management of the building itself. Also unlike cable systems, master antenna television is unregulated by both state and federal authorities.

For a new system, MATV is significantly cheaper than cable, and is often found in areas where cable systems politically or economically cannot survive. In particular, MATV systems are quite popular in large urban areas such as Chicago and New York City, excluding Manhattan.

MATV is seen as a threat to both the national cable program networks and the local cable operators.[17] The networks even consider MATV systems complying with their connection fees as grabbing the most attractive and lowest-cost installations in a locality. MATV systems also have the opportunity to pick and choose among cable networks, providing unwanted competition. While some major cable networks, including HBO, are refusing to provide authorized reception by MATV operators, unauthorized reception is unavoidable. At the same time, most MATV operators enter into an exclusive arrangement with their customers, preventing them from connecting with another local cable system, since the loss of even a few subscribers could be damaging.

Satellite networks have made the existence of all three competitors possible, and will almost certainly prevent any one from claiming a clear victory.

Low-Power Television

In the 1950s thousands of low-power television (LPTV) transmitters, called translators, sprang up in rural areas far from fully licensed stations; these translators amplified and rebroadcast distant signals on a different (translated) channel from the original so as to avoid interference.[18] Their coverage was limited to perhaps twenty miles, and such stations were not allowed to originate any programming.

In 1982 the FCC relaxed restrictions on LPTV, eliminating the limitation on rebroadcast of local signals and almost all ownership rules. Suddenly, this almost moribund part of the broadcasting industry received 12,000 applications for new licenses in five months, creating such a backlog that processing was by lottery.[19] The sudden attention was due to the fact that the new rules now allowed satellite networks to solve LPTV's two main problems: little choice and small audience.

An LPTV translator attached to a TVRO earth station has the entire range of national broadcasting to select from, not just nearby signals, and is that much more attractive to local residents. (The FCC ruling had actually been intended to encourage locally originated programming but, just as with the localism doctrine thirty years earlier, this hope was rarely realized.) LPTV had also suffered from its strictly limited audience, and its reduced broadcast area as a function of transmitting power remains unchanged. With joint ownership and control rules lifted, however, satellite systems allow the inexpensive creation of networks of LPTV stations with as large a receiving audience as possible.

Aside from transmitting power, the only other restraint on LPTV transmitters is the requirement that they do not interfere with other primary local signals, and, conversely, that these signals are allowed to interfere with their transmissions. Even so, with the advantages of satellite technology, they have the opportunity to preempt cable development in both rural and urban areas, and to create dozens, if not hundreds, of specialized mini-networks.

Private Satellite Dishes

The phenomenon of private—backyard—satellite dishes goes one step farther by placing these choices directly in the viewer's hands. Since legal restrictions on TVRO earth stations were eliminated in 1978, their growth has been prodigious. More than a million such dishes are in operation, and their installation and maintenance are a thriving secondary industry.[20]

Like a commercial TVRO earth station, home dishes require a powerful amplifier to enhance the attenuated satellite signal. Unlike commercial systems, however, which almost always are trained on a single satellite and maximize the quality of its reception, private dishes gain their value from being able to move from one satellite to the next, picking up all possible signals. The machinery that controls the orientation of the dish antenna is the most important part of any home TVRO installation; some of the latest systems allow the user simply to enter the name of the satellite in which he is interested into a control console that contains hard-wired tracking information.[21]

Private satellite dishes, in theory, allow users to receive all satellite broadcasts; not just the premium pay services, or even the national network programming, but also dozens of regular transmissions not intended for reception, typically, parallel coverage by several cameras during newscasts or sporting events. At the same time, the heyday of the private dish movement may have already passed, with satellite operators' decision to scramble both premium and normal broadcasts,[22] at least until the next generation of unscrambling technology.

Direct Broadcasting Satellites

Direct-broadcast satellites (DBS) go one step farther toward eliminating the formal downlink. "Direct" is a slight misnomer, conjuring up satellite transmissions received in a user's home without any intervening technology. Plans for DBS reception still require a small dish—two to three feet in diameter, as opposed to the ten to fifteen feet of backyard dishes. The DBS signal itself is so much more powerful than conventional signals, up to 100 times stronger, that no ground amplification is needed; since the DBS antenna is attuned to only the DBS satellite, no complicated tracking machinery is required either.

The Canadian government again was first to set up a DBS system (experimental in 1976, operational in 1980), which is mostly in use in the far northern territories. Japan also launched a trial DBS satellite in 1978, with its first regularly scheduled programming network set up in 1984. Many European systems have been proposed, but are foundering under political problems. In 1982 the FCC invited applications for DBS satellites and was immediately deluged with nearly thirty proposals. At that time, DBS was clearly thought to be the wave of the future, supplanting existing cable systems by the end of the decade, and bringing broadcasts to areas unserved and unservable by cable. Today, the dream of DBS, first proposed in the early 1960s, seems as far off as ever for at least three reasons.

First, the very power of the signal required for broadcasting creates several problems. Very powerful signals are that much more likely to interfere with one another, so that the limit on DBS programming from a single source is probably six to eight channels, not an attractive range of choices to compete with cable systems. The high power requirement also means a smaller reception area or footprint, so that the cost of a national network is substantially higher. The stronger signal is also more readily interfered with by atmospheric conditions.

Second, the initial burst of excitement by domestic DBS applicants was soon countered by mutual apprehension of the number of customers. When the Japanese system failed in 1984, only a few weeks after being put into orbit, the largest potential entrants (ComSat, CBS, and Western Union) postponed or eliminated their DBS plans.[23]

Third, there is considerable resentment surrounding the degree of control exercised by DBS. For the viewer in the United States, DBS represents potential savings at the cost of the personal freedom of a private dish or the wide selection of cable. All other domestic satellite systems see DBS as a particularly hostile competitor since it destroys the need for their distribution systems altogether. Internationally, these systems are government entities with even more cause to fear loss of control over what their nationals are viewing and, in this arena, political rather than technical obstacles seem insurmountable.[24]

ALTERNATIVES TO SATELLITE BROADCASTING

Local relay of broadcasting signals from microwave tower to microwave tower over AT&T's terrestrial network was, only a decade ago, the most common means of transmission. Now almost all signals, except for a few small private networks, are carried by satellite and then rebroadcast locally, sent out over cable or received directly into a viewer's home system.

Two other technologies have sometimes been proposed as alternatives to satellite broadcasting, although strictly speaking they should be thought of as adjuncts instead, and hybrid systems are quite possible. These are fiberoptics and multipoint distribution service.

Fiberoptics

Fiberoptics is the shorthand name for transmission of signals as a laser-generated beam of light at a particular frequency over very thin glass fibers. As noted in the last chapter, fiberoptics can be quite competitive with satellites in the case of one-to-one information transfer, but their advantages there evaporate for one-to-many information transfer, at least in today's market.

Fiberoptics' very capacity is its downfall. A single fiberoptics system, in theory, could carry up to a billion video channels.[25] Current systems can transmit on the order of 500 video channels, which is roughly equivalent to 1 million telephone calls. A million phone calls simultaneously on a New York-to-Washington fiberoptics network is a quite reasonable use. On the other hand, the total number of video channels in the world is closer to 100 than 500, and no one system uses more than a small fraction of these. Using fiberoptics to deliver television broadcasting would be enormously wasteful of resources, more than enough to offset possible cost advantages. Future systems offering additional data and two-way services ultimately might be able to benefit from fiberoptics, but if such systems come to exist at all, they are more likely to be extensions of telephony networks than television ones.

In a hybrid system, fiberoptics might carry a variety of signals on a high-traffic route between two centralized transmission or reception locations in order to emulate its one-to-one information transfer advantages. A hookup to individual installations would, again, be enormously wasteful of fiberoptics capacity, since the same signals would be going out simultaneously over the whole network. Again, piggybacking on the telephone system just to soak up extra capacity may be the exception to this.

Multipoint Distribution Service

Multipoint distribution service (MDS) is the transmission of high-frequency radio signals from an omni-directional microwave antenna. When originally proposed by the FCC in 1962, the designated channels for this service were too narrow to carry a video signal. In 1970 the restrictions were loosened a bit to allow short-range video transmissions, typically in the form of closed-circuit television for private business and educational systems. In 1974 the final step was taken, and MDS was recognized as an alternative form of television broadcasting with two channels allocated to each of fifty major cities.[26]

To receive an MDS signal, a special antenna is required that must be within line of sight of the original transmission. Most MDS programming is a rebroadcast of premium services like HBO, and thus MDS's largest problem is signal piracy. In a cable system, some physical connection with the system is required to obtain the signal; for MDS, sometimes called wireless cable,[27] interception by an unauthorized antenna is almost impossible to prevent.

Growth of MDS systems has long been retarded by the small number of channels available. In 1983 the allotment was changed to ten per city, which makes the possibility of hybrid satellite-MDS systems much more practical.[28]

OTHER TYPES OF BROADCASTING CONTENT

This discussion of the use of satellite systems in one-to-many information transfer has dealt almost exclusively with commercial television broadcasting. Obviously, the same considerations apply to noncommercial broadcasting. Satellite technology has widened the reach of educational and other public-interest programming even more than commercial broadcasts because of the informal systems it so easily permits. In a much narrower application, satellite networks are also used to broadcast navigation information, primarily, but not exclusively, for military traffic.[29]

Radio made its first use of satellite networks a few years before television. In 1978 the Mutual Broadcasting Company began operating a national radio network, followed the next year by the news stations of the Associated Press (AP) and United Press International (UPI). The three national television networks initiated their radio broadcasts by satellite in the early 1980s.[30] Today nearly forty radio networks are operating by satellite. Radio signals are also used to accompany television broadcasts in order to generate a stereo signal.[31]

Teletext

The newest form of one-to-many information is teletext: the one-way video transmission of text. Teletext, the "electronic newspaper of the air," uses the fact that not all the horizontal lines in a standard television transmission carry visual information. Approximately 20 of the 525 lines are used to determine the vertical blanking interval (VBI), the black band that can be seen at the edges of the screen when the picture is out of adjustment.[32] A data subcarrier can fill these lines with information that a special decoder in the viewer's home can recognize and display as text.

The first teletext service, CEEFAX, was offered by the British Broadcasting Corporation (BBC) in 1974. The service was government supported and was presented in a newspaper format. Rather than dialing channels, a viewer could choose from 800 pages of information under nearly continuous update. The BBC's commercial counterpart, the Independent Broadcasting Authority (IBA) soon started a nearly parallel advertising-supported service called Oracle. Teletext was carried one step farther by the British Post Office, whose Viewdata service operated more as a reference library, though one with a painfully small range.[33]

A French system, Antiope, and a Canadian, Telidion, soon followed. Although explicitly encouraged by the FCC in 1983, teletext, even in experimental form, has been slow to take off in this country, where there is no direct government subsidy of this service. CBS instituted a trial teletext service, Extravision (modeled on the French system), in Los Angeles in 1983. In Chicago, WFLD-TV has been broadcasting an active (one that allows a query of a particular page) teletext service, Keyfax, during the day and a passive (just a continuous roll through the pages) one, Nite-Owl, in the late-night hours.[34] Both experiments have attracted fairly modest audiences, and the offerings have usually been limited to UPI and Reuters newscasts, with occasional stock quote services.[35]

The resistance to public acceptance of teletext in the United States seems to be due to the embedded perception of television as an entertainment medium rather than an informational one, the reluctance to read a printed text on a television screen, and the relatively high costs and low capacity of teletext compared with its principal competition, newspapers.[36] Its only advantage, that of continuous updating, seems to be a fairly slim one.

TYPICAL APPLICATIONS

The range of typical applications of satellite information networks in one-to-many information transfer can be suggested by an overview of some approximate costs followed by three brief case studies.

Costs

The use of satellite information networks and the other technological changes described in this chapter have made the costs of one-to-many information transfer especially reasonable. One general benchmark is the cost of transmitting a video signal. Lease of a satellite transponder for a year costs roughly $1 million, or about $100,000 a month. Incidental use runs about $1,000 per hour. All of these figures are approximately one-tenth the costs of terrestrial microwave propagation, and all can be cut by a factor of four again if the signal is preemptible; that is, if the signal can be interrupted in favor of another transmission in the case of satellite difficulties.[36]

The earth station needed for a full-scale send and receive video broadcast costs between $100,000 and $500,000, a TVRO earth station about $10,000 (down from $150,000 in 1976 for comparison purposes); a turnkey home system about $3,000; and a private dish from a kit as low as $200 to $3,000 for the enterprising handyman.[37] It should be noted that the advances in satellite technology have brought down transmission costs by a factor of 10, but reception costs by a factor of between 100 and 1,000. The power behind these advances has clearly been a decentralizing, rather than a centralizing, force.

The same conclusion is true for the distribution technology as well. The last recorded transfer of a nonnetwork television station license in New York City was in 1976, for $100,000,000, which is about the same for operating a national pay television network today. The budget of a smaller private network or a local network station will be roughly one-quarter this amount.[38] In contrast, an LPTV or MDS system will cost well under $1 million to set up and operate, and a 400-unit MATV system considerably below $50,000.[39] Localism has the opportunity to prevail in ways unintended and uncontrollable by the FCC.

Superstations

One of the many unanticipated consequences of satellite broadcasting has been the creation of so-called superstations. Superstations are local, nonnetwork-affiliated television stations that have deliberately set out to send their broadcasts to distant cable systems in order to create a national base for their advertising and, not incidentally, to fill the always open need of cable systems for more programming.

The first superstation was Ted Turner's WTBS-TV (then WTCG), an independent station in Atlanta. Turner was fascinated by the success of HBO's satellite venture and had two special assets of his own: ownership of several sports teams and a large collection of old movies. In 1976 Turner formed a subsidiary, Southern Satellite Systems, to lease an RCA Satcom I channel. To attract subscribers, he offered his signal for a nominal fee to any cable system with an existing earth station, and promised to help finance earth station construction for those without.[40]

This foot in the front door has made WTBS one of the top five most watched satellite-delivered cable channels in the country, with an estimated audience of thirty million viewers.[41] The station's phenomenal growth led Turner to create a second national channel from scratch, the Cable News Network (CNN). WTBS has also been carrying passive teletext news services.

In turn, the success of the Atlanta venture inspired other local stations to go national. Following in Turner's footsteps are WOR-TV, New York City's channel 9, broadcasting on Westar 5; WGN-TV, channel 9 in Chicago, operating on Satcom-3; WPIX-TV, channel 11 in New York City, operating on Satcom-4; KTVT-TV, from Dallas, on Comstar 4; and XEW-TV, from Mexico City, providing Spanish language programming on Westar 4.[42] LPTV systems are also expected to make extensive use of superstation arrangements.

Hotels

Nearly all hotels have internal MATV systems for distributing signals to guest rooms. Sometimes these systems distribute signals from videocassettes in the basement or a local MDS carrier, but more often from their own TVRO receiving satellite signals.[43] These services may be on a pay-per-view basis for premium movies, or bundled into the room price.

The next logical step was for a hotel chain to link its TVRO station together to form a network of its own. Doing so would provide opportunities for volume discounts and other efficiencies. More important, a hotel network would also be able to provide private broadcasts of sports and other entertainment events from one of its locations and move into a gray area separating one-to-many and one-to-one information transfer by supporting video conferencing.

In 1983 the Holiday Inn chain created just such a network. Hi-Net Communications, a Holiday Inn subsidiary, in a joint venture with ComSat linked its 1,500 locations together by the Satcom 1R satellite.[44] The effort has been quite successful, and the Hilton, Hyatt, and Marriott chains are proceeding in the same direction.[45]

Other Private Networks

The satellite revolution transformed many marginal broadcasting entities into national networks. There are now almost forty "fourth" networks after thirty years of failure to come up with a successful competitor to CBS, NBC, and ABC. Among the most successful, as a class, have been the religious stations.

The Christian Broadcasting Network (CBN) was created in 1960. In 1976 the station had only three affiliates and jumped on the satellite bandwagon. In one year, 400 new affiliates were added.[46] By 1983, Pat Robertson's CBN had nearly 4,000 affiliates reaching some 20,000,000 viewers with around-the-clock programming.[47] More than a dozen other religion-oriented stations are attempting to match this success story.[48]

The largest single private network in terms of viewers is the Spanish International Network (SIN). SIN had been floundering for many years due to the high land line costs between its Spanish-language stations in a handful of large cities. It moved to Westar in 1977 and by 1983 enjoyed 200 affiliates with more than 25 million viewers.[49]

Other large private networks are dedicated to news or sports, the mainstays of the traditional national television networks. Only a few are devoted to broad-based public affairs and education programming. Instead, the growth of the private networks seems almost paradoxically rooted in their narrowness. Greatest common denominator programming needs are, in fact, well served by the big three television networks; satellite information systems allow the creation of a national audience for almost any specific interest.

CHAPTER 4
MANY-TO-ONE INFORMATION TRANSFER
DEFINITION

The third major use of satellite information systems is many-to-one information transfer; that is, the transmission of many signals simultaneously to a single receiver. The most familiar examples for the general public are satellite weather pictures. The generic activity of many-to-one satellite systems is called earth observation.

Many-to-one information transfer shares some of the features of the other two types of systems in addition to having distinct characteristics of its own. Like one-to-many information transfer, the communication is one-directional. For broadcasting, one-way transfer made the requirements for a formal downlink small; for earth observation, it is the uplink that is nonexistent in almost all cases. Many-to-one information transfer is closer to telephony in the number of different signals being transmitted over the network, however, and vastly exceeds one-to-one or one-to-many transfers in the variety of signals being processed.

For both broadcasting and telephony, geostationary orbit possesses special advantages in allowing coverage of a single, albeit large, area for many users simultaneously. The benefits of an earth-observation system lie in the ability to provide coverage of any area for a single user. Furthermore, for remote sensing and military purposes, a low earth orbit is required to obtain sufficiently accurate pictures. Also, in most cases, an earth-observation satellite is much more an active processor of signals than a mere relay device. Accordingly, such satellites require more power and have shorter working lives than those used for broadcasting and telephony.

HISTORICAL AND TECHNICAL BACKGROUND
Technical Limits

Earth-observation systems may be either active or passive. A passive system simply senses radiation being given off by objects on the ground; an active system generates a signal of its own and bounces it off the ground to be analyzed upon its return.

The most common type of active signal is a microwave transmission, more familiarly known by its World War II acronym, radar. The resolution of a radar signal, that is, the smallest object that can be reliably discriminated from its surroundings, is about 15 meters for commercially available systems, and much smaller but classified for military systems.[1] The radar image shows a ground object's approximate size and shape and, more important, its texture. Moisture level is especially well determined by radar imagery and, if the approximate nature of the ground cover is known, radar image sensing can be quite discriminating, that is, down to type and maturity of a grain crop. Another form of active sensing uses a laser signal. This signal is so sensitive and accurate, with a resolution of a small fraction of a meter, that it is virtually useless for any practical application other than precision altimetry for scientific experiments.

Active sensing has enormous advantages over passive sensing in that observation can be made by night or day, independent of cloud cover or local weather conditions. Its related disadvantages are also large. Relative power requirements are much higher, with an attendant rise in costs, and the size of such systems is more easily handled by aircraft than by satellites. While it is reasonably settled that passive sensing is not a violation of territorial sovereignty, active sensing is much more likely to be regarded as an incursion, and may even qualify as a trespass under United States domestic law.[2] Finally, active sensing systems cannot be differentiated, in theory, from weapons systems, and are feared for that reason as well.

Passive systems rely on naturally emitted ground radiation for their sensing. The radiation studied is usually in the range of visible light or the near-visible infrared. Simple photography was the original means of such sensing. A resolution of 30 meters for commercial systems is possible, with that of military systems many times greater. Its drawbacks are the separate development stage needed for processing a photographic image and the simultaneous representation of all parts of the visible light spectrum. Current systems break the ground radiation into separate pieces, or bands, such that each is sensitive not only to color but texture as well, and, analyzed collectively, can reveal much more than a conventional photograph.

The two available commercial sensing equipment systems are the Multispectral Scanner (MSS) and the Thematic Mapper (TM). The MSS system has provided all regularly available sensing images from satellites launched before 1986. It separates ground radiation into four bands[3]:

1. Band 4, covering a wavelength of 0.5 to 0.6 micrometers, captures blue-green radiation and is most useful for studying characteristics of water bodies, including depth, turbidity, pollution, and plant life, but reveals topographic features poorly.

2. Band 5, 0.6 to 0.7 micrometers, captures red radiation, distinguishing topographic and man-made structures well; it is best suited for studying land use patterns.

3. Band 6, 0.7 to 0.8 micrometers, filters the red and near infrared wavelengths to image vegetation patterns, gross landforms, and land-water boundaries.

4. Band 7, 0.8 to 1.1 micrometers, receives in the infrared ranges and so is best at penetrating cloud cover; it is most successful at tracing water system patterns and their effects on vegetation.

The resolution of the MSS is about 80 meters in a typical frame 185 kilometers square.

The TM has been operated on an experimental basis by the most recent United States satellite, but is commercially available only on the French SPOT system launched in February 1986. It divides ground radiation into seven bands[4]:

1. Band 1, 0.45 to 0.52 micrometers, captures blue-green light and is used to map coastal water and detect differences in soil and vegetation.

2. Band 2, 0.52 to 0.60 micrometers, is also in the blue-green range and is more accurate for measurements in deeper waters as well as for determining rates of plant growth.

3. Band 3, 0.63 to 0.69 micrometers, shifts into the red range and is used for classifying crops through measuring chlorophyll absorption, detecting iron ore, and mapping snow and ice.

4. Band 4, 0.76 to 0.90 micrometers, is still in the red range and delineates land-water boundaries and measures biomass generally.

5. Band 5, 1.55 to 1.75 micrometers, tracks infrared radiation

and monitors plant health through estimating water content.

6.   Band 6, 2.08 to 2.35 micrometers, is higher in the infrared range and is used to measure water temperature as well as mineral presence, and to identify soil.

7.   Band 7, 10.4 to 12.5 micrometers, is at the end of the infrared range and is best suited for general thermal mapping in applications such as plant health, pollution, and land use sensing.

The resolution of the TM is about 30 meters (except for the high thermal band, where it is about 120 meters) in a typical frame 80 meters square.

One differentiating aspect of earth observation systems is that the data they provide are so complex and voluminous that they are practically useless in their raw form. The most important parties in the industries are, in many ways, not the two giant systems operators, but the secondary tier of so-called value-added companies that process and interpret the remote sensing data for specific uses.[5]

Remote Sensing Satellites

Remote sensing in space goes back as far as 1947, when an experimental United States rocket carried photographic equipment into the upper atmosphere.[6] Earth photography was an important activity in the manned flights of the 1960s, but it was not until July 1972 that the Earth Resources Technology Satellite (ERTS, shortly thereafter renamed LANDSAT 1) was launched as a dedicated earth observation system. Following were LANDSAT 2 (January 1975), 3 (March 1978), 4 (July 1982), and 5 (March 1984).[7] All were plagued by technical problems, with LANDSAT 4, in particular, losing almost half its power immediately after launch. Another United States venture, Seasat, was a wide-resolution (1 kilometer), active, sensing oceanographic satellite that failed after three months in orbit.[8] Prior to SPOT's launch, the only other attempts at remote sensing were two Indian launches in 1979 and 1981. In the first, the sensing equipment never became operative; the second, dedicated to television pictures only, lasted for a few months.[9]

Despite these problems and initial over-optimism as to the value of remote sensing satellites, both their real and perceived contribution, was large enough to make "sensed" countries apprehensive of the quantity and quality of information being captured about their own territories. From the very start, the United States attempted to mitigate those fears by making the LANDSAT program part of its open-skies policy. Under this policy, all countries would have access, on a nondiscriminatory basis, to LANDSAT data essentially at cost, and the United States would assist other nations in constructing ground stations for receiving and processing LANDSAT satellite signals. Today, some ten countries have such ground stations; the majority are in South America, although Italian and Swedish stations receive data for use in a European Space Agency network.[10]

These tensions—increasing costs, sovereign sensitivities, and a commitment to international access—led to the only public refusal of a shuttle payload in the Spring of 1984.[11] SPARXX was a venture intended to launch a remote sensing satellite designed and built by Messerschmidt-Boelkow-Glohm GmbH (MBB) in the Federal Republic of Germany. ComSat and a German insurance company were partners in the venture until pressure by the United States caused ComSat to withdraw. SPARXX desired to keep proprietary rights to its sensed data for a period of time to maximize its return before releasing it on a first-come, first-served basis. SPARXX agreed to initial American stipulations concerning grandfathering existing data arrangements and assistance to the ground station network. Through NASA, the United States added more and more restrictions on the payload contract, including United States approval of all foreign contracts, access to all software, free use by this country of all SPARXX data, and free use by everyone of all data after ten years, until the payload was turned down because of SPARXX's refusal to accede. SPARXX has since turned to Arianespace for a launch slot; negotiations there have gone very slowly because of Arianespace's commitment to the French system, SPOT, launched in February 1986 and now the only operating competitor to LANDSAT.

The United States also had its own competitor to LANDSAT in the wings. The decision to control privately the nation's remote sensing system began in the Carter administration; moving the system from the space program to the National Oceanic and Atmospheric Administration in 1979 was seen as the first step in government divestiture. In addition to the cost and international pressure issues, the Reagan administration was also motivated by a broad philosophy of private control when, in 1981, it solicited bids from industry for a takeover of the LANDSAT system. Government conditions on the takeover were strong and funding was progressively tighter year after year; ultimately all the initial bidders dropped out. Finally, after intense government lobbying, RCA and Hughes Aircraft, designers and manufacturers of the current system, reluctantly agreed to form a joint venture, called Eosat, to carry on the program.[12] No sooner was the initial agreement made in 1984 than Congress cut all funding for the venture.

Additional budget cuts and the Gramm-Rudman Act have kept Eosat in limbo. At the present, LANDSAT 5 is expected to die in February 1987, with no possible plans for LANDSAT 6 until late 1988, which will leave an unfortunate gap in United States sensing capability.

Meteorological Satellites

One of the first satellites launched by the United States was a meteorological satellite: the Television and Infrared Observation Satellite (TIROS) in April 1960. Some forty short-lived weather satellites have been sent into orbit since that time by the United States. The present system is based on the NOAA (National Oceanic and Atmospheric Administration) satellites (NOAA 6 and 7, which circle the earth every 102 minutes, and an older Geostationary Operational Environmental Satellite (GOES 6) covering the Western hemisphere. (GOES 5, providing supplemental coverage, failed in 1985.) The Soviet weather satellite, Meteor, is in polar orbit. The other existing national systems—the European Space Agency's Meteosat, India's Insat, and Japan's Geostationary Meterological Satellite (GMS)—are all in geostationary orbit.[13] The very low resolution needed for weather monitoring—about 1 kilometer—allows the advantages of geostationary orbit not generally available to earth-observation systems to be used here.

This gross level of sensing has avoided the resentment that the LANDSAT-type systems generated. It also means that processing and interpreting the sensed data are that much simpler, with the result that roughly 125 countries have developed receiving stations for weather satellite data. The United States system also provides information to the World Meterological Organization (WMO) and, in turn, through WMO, receives local weather data from most of the rest of the world.

This reciprocal relationship, in which the United States in fact receives a greater volume of information than it provides, led Congress to pass legislation[14] prohibiting private control of the weather satellite system, which the Reagan administration desired as an adjunct to its LANDSAT commercialization initiatives. Nevertheless, the budget for meterological satellites has been cut sharply as well, and one current proposal is for LANDSAT 6 to do double duty as a weather and remote sensing satellite.[15]

Military Satellites

The number, much less the purpose, of military satellites in orbit is highly classified. It is estimated that somewhere between one-fourth and one-third of all satellites launched by the United States have been for military purposes, with a higher figure for the Soviet Union. The People's Republic of China is believed to be the only other nation to have launched a military observation satellite.[16]

Early military satellites were limited to using photography as a means of sensing, and had to deliver the actual film back to the surface. Today's systems are much more discriminating and sophisticated than the commercial systems described so far, with probable ground discrimination on the order of a small fraction of a meter. The dramatic uses of military sensing systems include early warning systems, surveillance, and reconnaissance (i.e., espionage), including not only ground observation but also interception of communications signals. The legality of spy satellites under international law has long been settled, and the Salt I (1972) and Salt II (1979) agreements[17] provided special protection for surveillance satellites, included under the euphemism "national technical means of verification." Nonetheless, even the existence of surveillance satellites was not formally acknowledged by the United States until 1978.[18]

Given the secrecy surrounding them and the nature of their applications, why should military observation satellites receive even a brief discussion in a treatment of commercial systems? The most obvious reason is that the nondramatic uses of military systems are analogs of civilian systems, just as they were in the one-to-one (telephony) and one-to-many (broadcasting) information transfer areas. Large parts of the activity of military sensing systems are dedicated to weather observation, geodesy, and position finding. It is also true that the commercial technology of today always turns out to be the military technology of yesterday. The congressional Office of Technology Assessment has noted the following ways in which civilian systems actually support military ones[19]:

- Providing complementary (and often unique) data on non-sensitive matters;
- Acting as a backup system for data and its processing in time of emergency or overload;
- Maintaining a broader technical base of personnel; and
- Providing other political and intelligence advantages, includ-

ing a means of releasing otherwise classified information, obtaining reciprocal information rights from other nations giving access for foreign sensing technology, and deflecting criticism of the United States through open distribution of data.

While not plagued by the budget cuts affecting civilian remote sensing and weather satellites, military observation systems have been curtailed by the sequence of launch disasters in the first half of 1986, which were especially hurtful to short-lived surveillance and espionage satellites.[20]

TELEMETRY AND OTHER COMMERCIAL ISSUES OF REMOTE SENSING

Stock Variables: Geological Structures, Metals, and Minerals

Resources can be thought of as either flow variables, varying over time and inexhaustable, given proper management, such as agricultural crops, or stock variables, fixed for all time and therefore inexorably exhaustable, even given careful management, such as gold mines. Stock variables are obvious candidates for exploitation by remote sensing systems because of their inherent value and stable location with respect to time and space.

Accordingly, by far the largest private user of remote sensing information is the petroleum industry, which purchases most of its data from a value-added company that caters to oil and mineral companies, called Geosat.[21] For obvious reasons, the details of these ventures remain quiet for domestic ventures and tend to become public only when international negotiations are necessary.[22]

Although it serves to focus efforts quite well, remote sensing of oil reserves is only a suggested first step at best, requiring closer aerial surveys and, ultimately, field explorations for only a modest prospect of success. Uranium deposits have been located by remote sensing,[23] and a discovery of copper resources in Pakistan is especially well documented.[24]

Flow Variables:
Agriculture, Forestry, and Wildlife

The largest public user of remote sensing data is the U.S. Department of Agriculture. Its Agriculture and Resources Inventory Surveys through Aerospace Remote Sensing (AGRISTARS) program, founded in 1980, performs the following functions with respect to barley, corn, cotton, rice, soybeans, and wheat, on a global basis[25]:

- Assessing early warning/crop condition;
- Forecasting foreign commodity production;
- Forecasting potential crop yield by region;
- Estimating soil moisture;
- Inventorying renewable resources, including forests and range land; and
- Monitoring conservation and pollution;
- Agricultural research; and
- Census of domestic crop acreage and land coverage (in conjunction with other surface-measurement systems)

Through these activities, the effects of stress, disease, weather conditions, and irrigation are measured and analyzed for users, who are primarily commodities traders and international farming interests.

Most remote sensing of forests is made by aerial surveys rather than by satellite due to the inability of the latter to penetrate crown cover, for the most part. Where all the trees are of a given type (e.g., a rubber plantation), satellite observation can provide the same estimates of health as for ground vegetation, and is always useful for gross estimates of standing timber.[26] Wide-scale environmental effects, such as fire, frost, and pollution, are also readily tracked by satellite systems.[27]

Aerial surveys are almost exclusively used for analyzing patterns of wildlife migration with one large exception; satellites are much more cost effective for tracking schools of fish. By identifying likely feeding areas, such systems have been used to improve the catches on the West Coast.[28] This is also the primary use intended for the proposed Japanese system.

Cartography and Geodesy

To be mapped accurately, points on the earth's surface must be within sight of one another; one consequence of this fact is that for 200 years the face of earth's moon was more precisely known than the face of the earth, with most ocean and many desert areas being simply unknown.[29] This most forthright use of earth observation satellites has been to provide the first truly accurate maps of the planet.

In less broad terms, images from the LANDSAT system have frequently been used by local government planners to design or evaluate zoning schemes and to inform other aspects of land use planning. Urban sprawl, degree and rate of industrialization, and traffic patterns are the most common applications.[30]

METEOROLOGY AND OTHER SCIENTIFIC USES

Weather Prediction, Meteorology, and Climatology

The ability of satellite sensing systems to make accurate maps is linked to the improved weather forecasts they allow: the ability to observe the oceans, or more exactly, ocean currents, whose temperature and direction determine most of the world's weather. Thanks to weather satellites, five-day forecasts today are as good as two-day forecasts in the 1970s.[31]

Longer-term forecasts are still basically statistical but are being enhanced in conjunction with ground station data.[32] In this middle range, satellite sensing systems are estimating more general trends as part of meterological studies, rather than making specific predictions.[33] The most significant benefit weather satellites provides is in their long-range climatological analyses, studying such topics as ozone concentration in the atmosphere, fluxes in solar radiation, dispersion patterns of contaminants, and the alleged greenhouse effect of a warming planet.[34]

The NOAA system of weather satellites makes more than 20,000 observations of weather patterns daily. Signals from these satellites are easily processed to offer a televisionlike image to the general public. This civilian system is closely matched by the Defense Meteorological Satellite Program (DMSP), which operates an analagous network solely for the use of United States armed forces.[35]

Hydrology and Oceanography

The study of water resources by satellite observation systems falls some-where in between remote sensing for mineral deposits and weather forecasting applications. Since such resources are not typically appro-priable by private parties, national planners and broad farming interest are the typical users. Common applications include the following[36]:

- Estimating surface and ground water requirements for use in managing reservoir runoff and irrigation systems;
- Flood protection and control;
- Designing drainage projects;
- Measuring water quality; and
- Estimating sedimentation and erosion.

Satellite images have also been used to select hydroelectric power sites in Scandinavia and new villages in Pakistan.[37]

Satellite systems are still a relatively poor means of identifying ocean resources, except in shallow coastal waters. Their contribution to oceanography lies chiefly in studies of currents done for meteorological purposes, as discussed above. As another intermediate application, United States satellites also provide a continuous glacier watch in the North Atlantic as a service to shipping.

Archaeology and History

Archaeology from the air goes back to ballooning days. Even Charles Lindbergh used the airplane for archaeological reconnaissance in the Caribbean.[38]

A secondary scientific use of satellite observation systems is in identifying extinct communities and buried structures in a way unavail-able to other types of survey. Satellites have been used to discover lost ruins in Guatemala and Peru by observing patternlike anomalies in the vegetation covering these areas. Their most successful application in this regard has been the revelation of probable Stone Age settlements and underground river systems in the Sahara Desert.[39] The dry desert sands allowed active radar sensing to detect structures as much as 5 meters deep that are thought to be communities on the banks of rivers that disappeared more than 5,000 years ago.

In more down-to-earth applications, satellites have been employed in this context to identify and bound areas receiving federal protection under various pieces of preservation legislation.[40]

OTHER SOCIAL AND POLITICAL USES

Law Enforcement

Satellite observation systems can measure ground conditions over large and often inaccessible geographical areas, and images are usually available with sufficient regularity for transient effects to be caught. This ability resulted in the LANDSAT network being used frequently to assist federal law-enforcement agencies. Most of these cases involved environmental pollution and were brought to court during the 1970s.[41] Satellite sensing has been able to provide dramatic evidence of fish kills, oil spills, strip mining, toxic waste dumping, and other chemical discharges. This evidence has also been frequently analyzed by defendants in attempts to prove their innocence. Hampering the efforts of both sides is the fact that both NASA[42] and NOAA, its successor in operating the LANDSAT network, have declined requests to make special runs over or analyses of any particular area.

LANDSAT images have also been used in interstate litigation in similar applications, even in a suit brought by Mexico against the United States.[43] No satellite images have been used in any criminal proceedings, however, given the higher standard of proof required there. It seems to be the law, however, that satellite observation does not represent an unconstitutional search.[44]

Distress Signals and Position Finding

In the 1970s the Soviet Union developed a Search and Rescue Satellite (SARSAT) network to pick up and transmit distress signals from ships in its merchant marine. The United States joined the system shortly thereafter, and it became a worldwide operation covering signals from land as well as sea. The SARSAT signal is generated by an emergency beacon that is activated manually, or by being turned upside down or immersed in salt water.[45]

Several hundred rescues have been credited to the SARSAT network, about equally divided between downed aircraft and ships at sea.

Although the United States briefly withdrew from the system in 1984 due to budgetary pressures, the current arrangement now consists of two Soviet, two American, one French, and one Canadian satellite.[46]

A related system is the Transit satellite network, operated by the U.S. Navy but open to all civilian users. It locates an object at sea to within a few hundred yards. Transit is a navigation rather than emergency system. It is currently being updated by the NAVSTAR Global Positioning System (GPS), a group of eighteen satellites intended to become operational by the end of the decade, which can measure location to within 50 yards at sea and within 3 yards at a major port.[47]

Chrysler Motors has plans to put access to GPS as an option in its luxury-model cars.[48] A similar location device is being marketed for personal use by a firm called Geostar. Originally intended as a personal distress signal for use within cities, it is now believed that its most profitable use may be in tracking stolen property.[49]

Disaster Prediction and Relief

Satellite observation systems may be used to analyze the risk of natural disasters, provide an early warning system for their occurrence, and assess the change that they cause. The LANDSAT Emergency Access and Products (LEAP) system is designed to provide images within hours of a request to public users, typically, other nations and, domestically, the National Guard and the Army Corps of Engineers.[50] Floods and oil spills are the most frequent applications, but earthquakes, landslides, and volcanoes have also been analyzed.[51] LANDSAT was even called in to confirm for the West the nature and the extent of the Chernobyl reactor fire in the Soviet Union in April 1986.[52] After the fact, satellite sensing systems have also been helpful in coordinating relief and reconstruction efforts, especially with respect to flood management.

MILITARY USES

Early Warning Systems

The least controversial and most public military use of observation satellites is as an early warning system to detect the launch of land- or submarine-based ballistic missiles. The blandly named Defense Support Program (DSP) consists of three satellites in geosynchronous orbit over the Atlantic, Pacific, and Indian Oceans.[53] DSP satellites identify launches through infrared sensing of the heat plume generated. The current generation of satellites can determine the trajectory of the satellite as well. Their related function of monitoring nuclear explosions, with respect to arms control and testing agreements, will ultimately become part of the GPS position-locating network.[54]

Surveillance

Surveillance satellite systems typically cover a fixed area of the surface or space from geostationary orbit. Their announced use is monitoring of arms control compliance on a deployment level, rather than explosion or launch. In this role they have the official sanction of international treaty, but their likely applications go much further.

Reconnaissance Satellites

Reconnaissance satellites are those that follow eccentric orbits that may come relatively close to the ground, within 100 kilometers, to obtain the clearest possible images, frequently with the aid of active sensing. Although not at all limited to sensing by camera, their activities are usually photographic reconnaissance. This is differentiated from electronic reconnaissance, which involves the active interception of communications signals by "ferret" satellites.[55]

The number and mission of such satellites is highly classified. They are managed by the National Reconnaissance Office (NRO), the very existence of which is still not officially acknowledged by the United States.[56] In a network of approximately sixty satellites, the successful current generation is the KH-11 (Keyhole) satellite, which has both active and passive sensing systems, undisclosed interception capabilities, and a large capacity for in-board processing and analysis of the data it receives.[57]

ALTERNATIVES TO SATELLITE OBSERVATION

For large or relatively inaccessible geographical areas, there are no alternatives to satellite observation because of physical constraints; for surveys of the territory of another nation, there are no alternatives because of political and legal constraints (as well as the practical ease with which balloons and airplanes can be shot down). For high resolution, independent of cloud cover for domestic or nearby areas, airplanes, balloons, and sounding rockets have some limited appeal.[58]

TYPICAL APPLICATIONS

The range of typical applications of satellite information networks for earth observation can be suggested by an overview of some approximate costs, followed by three brief case studies. The case studies are less relevant here than in earlier chapters, since they can only reflect experience under the governmental LANDSAT system, and not the private Eosat and SPOT ventures.

Costs

SPOT's original tariffs[59] for a single MSS frame are about $155 for a black-and-white print and $410 for a color print. TM frames run from $740 (black-and-white) to $790 (color). Computer compatible tapes (CCTs) range from $1,475 to $2,550.

Eosat's proposed prices have fluctuated through 1987, generally responding to SPOT initiatives. The main discrepancy is in the area of CCTs, with an Eosat MSS CCT costing only $660 but a TM tape running $3,300.[60]

Both firms have special priority and processing services. Eosat's running a fixed surcharge of 300 percent and SPOT's to be determined by negotiation.

Deforestation in Thailand

Anecdotal evidence led the Thai government in 1972 to request a remote sensing inventory of its forest cover.[61] The results showed that, at the current rate of deforestation, Thailand would lose all its forest cover within 150 years. An aggressive program of tree planting was coupled with a continuing sensing schedule that remaps the entire country every three years, with a systematic update every three months.[62]

Only satellite observation could have allowed cost-effective confirmation of the initial impressions, provided guidance to the reforestation efforts, and offered continuous monitoring of the success of these efforts.

Volcanoes

When Mount Saint Helens erupted in Washington in 1980 a GOES meteorological satellite began almost immediately to monitor the event at half-hourly intervals. Real-time measurement of the spread of volcanic dust was used to alert aircraft and down-wind localities. The eruption of El Chichon in Mexico in 1982 was an even more dramatic blast. It received worldwide monitoring by NOAA to measure its possible effects on long-term climatological changes. The progress of the dust cloud, even when seemingly diffused in the atmosphere, was tracked by measuring changes in sea-surface temperatures due to the interference of the particles with solar radiation.[63]

State Information Systems

Many states have made use of satellite sensing data for their own programs. Starting in November 1984, Arizona enforced its own water laws by satellite. Farmers who are illegally using excess water to irrigate their crops are being fined $10,000 a day for a quick payback to the state's $300,000 investment in image enhancement technology.[64]

Idaho, Washington, and Oregon have banded together as the Pacific Northwest Regional Commission to make use of LANDSAT data at savings of one-third to one-half over aerial surveys in such varied applications as agricultural/crop surveys, wildlife habitat inventories, identification of weed infestations, zoning analyses, and pollution studies.[65]

Part 2
Satellite Information Systems: Who Are the Players?

Chapter 5
Regulation

FORMS OF REGULATION

Regulation is the legal control of private enterprise by government. Such regulation may be either passive or active. Passive regulation concerns itself with minimal monitoring and list keeping, usually to provide a neutral basis for dispute adjudication, as in a requirement for registering deeds of real estate transfers, as well as safety standards. Active regulation may take the form of either constraint, punishing a particular activity, as by a fine, or incentive, promoting a particular activity, as through tax benefits. Active or passive, the forms regulation may take toward satellite information networks include regulating access, content, competition, ownership, and charges, and allocating orbits and frequencies.

Regulation of Access

Regulation of access is a threshold issue: if a satellite cannot be launched, obviously there is no system. Since the cost, risk, and national security consequences of rocket launches are so large, all effective space programs to date have been national efforts. Only two entities—NASA for the United States, and Arianespace, the quasi-private French corporation—have launched satellites for private interests. For many years, OTRAG, a private West German venture, has offered assistance in constructing launch sites without any completed projects. More recently, the People's Republic of China announced its intention to launch commercial satellites for hire, but as yet has not actually done so. Many other proposals are in the development stage.[1]

If an arm of the national government is the only means of

launching a satellite, then access is regulated directly. Under international law, such regulation may even be required. This direct regulation has been rarely exercised. Although NASA reserves the right to approve all payloads in advance, its inspection is typically limited to passive list keeping and safety standards functions. No payload for a United States party has ever been refused on the basis of its function, but at least one foreign project has been turned down.[2] Rather than outright refusal, however, private parties are at the mercy of NASA's initial scheduling proposals as well as actual launch delays, whether caused by last-minute technical problems or the need to bump a commercial payload for a military one. The recent space shuttle tragedy has greatly exacerbated such delays.

Arianespace is still at the stage of actively seeking customers rather than turning them down. Recent losses and delays, however, as well as traffic shifted from the space shuttle, have created critical delays there as well.

Regulation of access is also indirect to the extent that a national government defrays the cost of access. Charges for launch by both NASA and Arianespace are claimed to be considerably below actual costs.[3] Indirect regulation also takes the form of government support for research and development.[4] Exceptions to antitrust considerations have often been made by the United States to allow joint ventures of private competitors in satellite networks.[5] Also, while space commercialization remains a tax vacuum for most purposes, from the very start communications satellites have qualified for the investment tax credit by a special amendment to the Internal Revenue Code.[6] Through these and other sorts of subsidies, government regulation is actually used to promote the development of satellite networks. Regulation of access, in terms of restricting reception, is fortunately almost nonexistent in the West.

Regulation of Content

Despite the fact that not only launch facilities but also telecommunications facilities are government agencies, at least, outside the United States, the contents of satellite communications are regulated only minimally.

The content of one-to-one information transfer is not regulated except for local prohibitions against use of telephone lines for obscene or otherwise harassing calls. In fact, many seemingly criminal activities involving theft of data or services over telephone lines are not treated as crimes.[7] In both cases, there are no special "satellite" issues. These may arise, however, over international data transmissions in what is called the transborder data flow problem, where the country from which data are being transmitted claims proprietary rights to the information.[8]

Similarly, regulation of content of one-to-many information transfer is minimal other than areas such as libel, truth in advertising, and the political equal-time fairness doctrine, all of which have no special relevance for satellite networks per se. The controversial area here is again the international arena, where receiving, rather than broadcasting, states may wish to regulate the content of transmission from direct-broadcasting satellites.[9]

It is only with respect to many-to-one information transfer that regulation of content is actively exercised. When carried on by nations, it flirts with espionage; when carried on by private interests, it raises charges of exploitation of the observed country's resources.[10] The one shuttle payload refused was an earth-observation satellite system whose operators did not wish to comply with United States restrictions on content and publication.

Regulation of Competition

To the extent that satellite networks are state controlled, if not state operated, there is likely to be strong regulation of competing networks. Competing systems were first authorized in the United States in the early 1970s for domestic operations, but competing international systems, at least in terms of one-to-one information transfer, are quite strictly regulated.

Recent United States experience has taken two quite distinct paths: promoting competition with government systems and facilities, and allowing anticompetitive joint ventures in the private sector in order to provide such federal competition.[11] Still remaining, however, is the requirement that such competition be demonstrably in the national interest; for example, the level of competition is to be fine-tuned as an instrument of policy.

Domestic competition is quite active in one-to-one systems, and fairly so for one-to-many systems. For many-to-one systems, the United States program is the only operating candidate, although potential competitors are waiting in the wings.

Regulation of Ownership

Ownership of domestically operating networks for one-to-one information transfer is regulated indirectly through common carrier requirements. That is, corporations' ownership of systems open to the general
public is limited in that they must allow connection to the network
(with reimbursement) to all comers. In most other countries, there is no
private ownership of domestic one-to-one information transfer systems,
satellite or otherwise. Similarly, with the exception of the United States'
ComSat, all parties to international one-to-one satellite system operations are governments.

United States control of one-to-many information transfer networks is more tightly regulated as to both the initial grant of a broadcasting license and the number of different stations a network may
carry. As above, most foreign and international systems are state operated. Despite the United States' moves to operate privately the only public many-to-one network, it is an open question whether wholly private
actors can technically or legally operate such systems.

Regulation of Charges

The rates that network operators charge for system use may also be
subject to regulation. In the United States, private one-to-one networks
are unregulated in this fashion and, even for public networks, AT&T is
the only carrier whose rates are subject to governmental scrutiny on a
rate-of-return basis. Satellite systems are a concern to such an inspection only as one in a long list of cost elements. For international
one-to-one systems, however, satellite deployment and operations are
virtually the only costs incurred. These are also regulated on a rate-of-
return basis by an international cooperative.

Most one-to-many systems receive their financial support from
either advertising or subscription so that their charges are regulated by
the market rather than by the government. (In some cases, local government approval of a cable television franchise has turned on the minimum cost of subscription.) Up to the present, the United States has
insisted that the charges for use of these systems be placed on an incremental cost basis. This policy is intended both to defend a government
monopoly and to promote access as widely as possible.

Orbit Allocation

For satellite information systems, geostationary orbits are especially desirable for the coverage they can provide at minimum cost.[12] Competition for them is high. Currently, the orbits are given out on a first-come, first-served basis, with regulation being almost wholly passive. The regulation that does take place is the paradigm of the list-keeping approach and is performed by an international body, with which domestic regulators invariably agree. What enforcement capability does exist is solely at the national level.

Claims to ownership of a particular orbit, whether by prescriptive use or national sovereignty, have not been recognized, although interference with an existing system is prohibited by international law.[13] Currently, orbit allocation policies call for a minimal spacing between satellites, more to avoid overlapping of transmissions than physical collisions.

Frequency Allocation

Almost as important as a satellite's orbit are the frequencies at which it is allowed to broadcast, with specialized uses requiring higher frequencies.[14] Frequency allocation is administered by the same mechanisms as orbit allocation. Preventing interference is the primary goal.

OVERVIEW OF DOMESTIC REGULATION

Regulating Access in the United States

Although all United States satellite launches, as of early 1986, have been made on government vehicles at government facilities, it has been administration policy since 1982 to encourage private launches using expendable launch vehicles (ELV's)—that is, the familiar missiles of the 1960s and their immediate successors—as a supplement or an alternative to the federal manned space transportation system, the shuttle. Many such ventures are under development,[15] and the unavailability of the shuttle through 1988 should certainly accelerate their implementation.

The regulation of private launches explicitly follows the same guidelines that implicitly control governmental launches. Initially, proposals for private launches required approvals by the Federal Aviation Administration, FCC, and Departments of State, Defense, and Transportation, and the Coast Guard, in addition to the National Security Council, Office of Management and Budget, NASA, Office of Science and Technology Policy, the Cabinet Council on Commerce and Trade, and Senior Interagency Group for Space, among others.[16] This, of course, was in addition to the need for an arms dealer's license (for a missile), an export license (since the missile is being sent "out of the country"), and an import license (in case it returns).[17] On 4 July 1982 President Reagan issued a directive that national space policy was to "provide a climate conducive to expanded private sector investment and involvement in space activities."[18]

The first reaction of the private sector was a call to end the hodgepodge of regulation for launches. In February 1984 (coincidentally, just after the first satellite losses on the shuttle) the president designated the Department of Transportation (DOT) as the lead agency for coordinating private ELV ventures. DOT would arrange all necessary reviews by other agencies and would have the power to suspend existing regulation. In addition, DOT was charged to perform the following functions[19]:

- To act as a focal point within the federal government for private-sector space launch contacts related to commercial ELV operations.
- To promote and encourage commercial ELV operations in the same manner that other private United States commercial enterprises are promoted by United States agencies.
- To provide leadership in establishing, within affected departments and agencies, procedures that expedite the processing of private-sector requests to obtain licenses necessary for commercial ELV launches, and to establish and operate commercial launch ranges.
- To consult with other affected agencies to promote consistent application of ELV licensing requirements for the private sector, and assure fair and equitable treatment for all private-sector applicants.
- To serve as a single point of contact for collection and dissemination of documentation related to commercial ELV licensing applications.
- To make recommendations to affected agencies and, as appropriate, to the president, concerning administrative measures

to streamline federal government procedures for licensing commercial ELV activities.

- To identify federal statutes, treaties, regulations, and policies that may have an adverse impact on ELV commercialization efforts, and to recommend appropriate changes to affected agencies and, as appropriate, to the president.
- To conduct appropriate planning regarding long-term effects of federal activities related to ELV communication.

The DOT's responsibilities became formalized in the Commercial Space Launch Act of 1984.[20] This act granted the department regulatory power over any party intending to launch a vehicle or operate a launch site within United States territory, and over a United States citizen engaged in the same activities anywhere. These activities require the granting of a license by DOT contingent upon a review of the mission and launch safety, and other ancillary conditions.

The mission review component of the licensing process considers the impact of a launch on the international obligations of the United States (see Overview of International Regulations) and on national security and foreign policy issues as well.[21] Of primary concern is that the intended payload does not interfere with the activities of another space object. For satellite information networks, this means conformance with frequency and orbit allocations by the International Telecommunications Union. The bulk of the mission review is, in fact, carried on by the FCC and the Department of State.

The launch-safety review rules on site evaluation, range safety expertise, tracking and instrumentation procedures, including local frequencies temporarily allotted by the FCC for telemetry and control, and systems for aborting the flight in case of difficulties.[22] The review also considers the proposed vehicle design, with quick approval likely for missiles that have served as launchers in the past. Most of this review is, in fact, performed by NASA officials. Although NASA does not have any interest, responsibility, or authority for regulating private launches, it does possess almost all the relevant equipment and expertise for performing this role.[23] If both reviews are favorable, DOT oversees several final restrictions on the launch itself, including adherence to air-space restrictions, requirements for third-party liability insurance, and provisions for federal inspection.[24]

Except for the need for an import license, no existing regulations were actually removed by the Commercial Space Launch Act, even though some are quite outdated. For example, part of the pertinent FAA regulations, were originally designed to cover firecrackers and hobbyist rockets. The coordination role of DOT eases this burden somewhat. Also, the DOT secretary has the ability to waive any existing regulation as a license requirement if it "is not necessary to protect public health and safety, the safety of property, and national security and foreign policy interests of the United States."[25]

In both regulation of access directly (by ELVs or the shuttle) and indirectly (by government subsidies and other promotions), the policy of the United States has consistently been limited to minimal safety and list-keeping functions. Beyond the already thriving communications satellite industry, other aspects of space commercialization (especially materials processing) are expected to benefit from these generally de-regulatory trends.[26]

Other Forms of Regulation in the United States

The other principal actors in the regulatory arena of the United States are the FCC, NASA, and ComSat.

The FCC was created by the Communications Act of 1934 with the intention of seizing control of communications from the Interstate Commerce Commission.[27] Its seven commissioners regulate all United States interstate and international telecommunications, including telephony, radio, and television broadcasting.

The FCC's direct regulation of satellite information systems lies solely in its review of conformance with International Telecommunications Union assignments of frequency and orbit. Until 1982, however, the FCC also regulated charges for use of satellite transponders, where system operators were offering capacity to third parties.[28] This control was based on the argument that satellite communications capacity was a scarce resource, access to which had to be available on a reasonable and nondiscriminatory basis.

At least temporarily, worries over total satellite capacity have disappeared. "Scarce resource" regulation still holds, however, with respect to television broadcasting licenses, due to the limited number of channels available. These licenses, reviewed every three years, create an effective oligopoly in any broadcasting area. Because of this concentration of power, the FCC has held since 1971 that broadcasters cannot own cable television systems (although, in fact, the same corporate entities often own both, and this rule has been held open to change by the FCC itself) and, since 1975, that broadcasters cannot own a dominant newspaper within a broadcasting area as well.[29]

For more than ten years the FCC has been more in the business of deregulation then regulation. A list of FCC actions over that time indicates how few former controls still exist[30]:

- Competing domestic satellite systems are allowed (1972).
- Resale of communications services is permitted (1976).
- Licenses are not required for receive-only satellite earth stations (i.e., backyard dishes); AT&T satellites are no longer limited to voice communications; and a long-standing controversy over distinguishing communications (regulated) from computing (unregulated) disappears (1979).
- Only dominant carriers require detailed tariff regulation (1980).
- International carriers are allowed to compete in domestic markets (1981).
- ComSat may go into business for itself, rather than remain a carrier's carrier; voice/data distinctions are eliminated; enhanced services (those in which a signal is modified for some added value) are deregulated; applications for direct broadcasting satellites are encouraged (1982).
- Private satellite systems competing with Intelsat are encouraged (1983).

Finally, local regulation of satellite dishes cannot compromise communications competition, but must be reasonably and clearly related to "health, safety, or aesthetic" factors (1986).[31]

NASA was created by the National Aeronautics and Space Act of 1958.[32] Its formation represented President Eisenhower's deliberate decision to have the nation's space program developed under civilian rather than military authority. The NASA Act declared that "activities in space should be devoted to peaceful purposes for the benefit of all mankind," language that would be echoed in later international treaties. While NASA was given nonexclusive jurisdiction over civil space applications, the Department of Defense naturally kept its own weapons systems, communications networks, surveillance operations and research and development in support of these activities; in NASA's early years, the DOD even supplied special funding for NASA's own development work. The agency continues to provide flights for the military from time to time, but not without considerable friction: the Air Force has long asked for, and recently received, permission to operate its own shuttle flights. Until 1973 NASA was nominally run by a National Aeronautics and Space Council, whose members included the president and vice president; secretaries of state, defense, and transportation; and chairman of the Atomic Energy Commission. It is now an agency of the Department of Commerce.

At this time NASA has only one explicit regulatory role, that of technical adviser to ComSat. In practical terms, its power is quite large. Its pricing structure and commercialization initiatives effectively determine the world market for satellite networks. The agency has launched almost every Western satellite system, including all ComSat and Intelsat satellites. All international joint ventures of the United States have been carried on by NASA. In a consulting role, NASA has effective veto power over all private launches in the United States, and most of these ventures are, in fact, carried on by former agency employees. Finally, NASA is the chief supporter of communications and space vehicle research in the West. Although NASA's fortunes today are at an all-time low, given the Challenger tragedy, budget constraints, and charges of bureaucratic mismanagement, it will continue to provide effective leadership and control of satellite network development for the near future.

ComSat was created by the Communications Satellite Act of 1962.[33] As plans progressed for Telstar, the first nonmilitary communications satellite, there was considerable argument in the Kennedy administration as to whether satellite systems should be publicly or privately owned. The lack of direct national security involvement, the efficiency of private sector investment, a general bias against nationalization, including fevered charges of "creeping socialism," and the need for some intellectual property protection for system operators all suggested private ownership. Negotiating status in international ventures, accordance with the practice of other states, avoidance of private monopolies, the return of profits to the public, and the desire for complete coverage and nondiscriminatory access, even if unprofitable, militated for public ownership.

The final compromise between Kennedy, congress, and the industry players was the Communications Satellite Corporation, ComSat. ComSat was originally a strange hybrid, with ownership split fifty-fifty between communications common carriers—with AT&T having the lion's share—and public shareholders. The government retained the right to appoint a portion of the board of directors. By its own enabling legislation, ComSat was supposed to be regulated as to rates and access by the FCC, and to follow technical advice from NASA and policy advice from the Department of State. In particular, ComSat was specifically committed to deal with other countries to set up and operate a global communications system. Competitive bidding requirements and the mixed ownership were supposed to keep everything honest.

From the very start ComSat was troubled by three ambiguities. First, the role the common carriers were to play was quite unclear. For example, AT&T was at the same time a chief competitor to ComSat through its ocean cable systems, ComSat's largest client, the manufacturer of most of ComSat's equipment, and the owner and operator of the terrestrial network to which ComSat was tied.[34] These pressures, plus conflicting directives from the government agencies involved, led the common carriers to sell off all their interest in ComSat by the early 1970s.

Second, although the ComSat legislation seemed specifically to mandate a single system, this was not the preference of most of the participants. In 1972 the FCC authorized competing domestic systems, which naturally precipitated the dropout of the common carriers to set up their own networks. The status of competing international networks is currently a hotly contested issue.

Third, ComSat as a corporation for profit is fundamentally at odds with its role as United States representative. This tension has reduced, to some degree, as ComSat has moved from operator of the international system, Intelsat, to a mere, though still the largest, member. Also, in 1982 ComSat was authorized to go into business for itself rather than be constrained as a carriers' carrier. Nonetheless, ComSat sees itself now as the only one of many competing satellite system operators that is subject to a high level of government control.

Domestic Regulation in Other Western Countries

The extent of direct government regulation of telecommunications is much greater in other Western countries than in the United States. Typically, telecommunications are both provided and regulated by a government agency usually referred to as a PTT, for post, telephone, and telegraph, given their historical roots, but now including broadcasting as well. Most PTTs are associated with the manufacture of equipment as well, are reluctant to promote domestic competition or to deal with several foreign competitors, and use their profits to defray postal system expenses.[35]

These patterns are slowly changing. In the United Kingdom, post office functions have been split off from British Telecom, which is now 51 percent privately owned with at least one major competitor.[36] Japan is in the process of converting both its domestic and international telephone systems to private ownership.[37] The governing document of the European Economic Community, the Treaty of Rome, which discourages intra-EEC constraints on competition, has recently been held to apply to telecommunications.[38]

On the negative side, United States firms must still deal with foreign PTTs, for the most part, in order to be connected with other domestic systems, and most Western nations except Canada require licensing of receive-only earth stations. Movement in this area is characterized more by successive commissions on "the future of telecommunications" than by any realized deregulatory trend. In general, the most expeditious arrangements can be made with the handful of nations that have an organization, similar to ComSat at its outset, dedicated to satellite network operations: Canada (Telesat), Japan (NASDA), Italy (Telespazio), and India (India Space Research Organization, ISRO).[39]

OVERVIEW OF INTERNATIONAL REGULATIONS

International Treaties

The United States is a party to four multilateral treaties that have a general bearing on satellite information networks.

The first, the Treaty on Principles Governing the Activities of States in the Exploration and Use of Outer Space, Including the Moon and Other Celestial Bodies (Outer Space Treaty, or OST), became operative in 1967.[40] It states in Article II that no part of space is "subject to national appropriation by claim of sovereignty, by means of use or occupation, or by any other means." That is, a nation cannot *own* an orbit. At the same time, interference with another's peaceful activities in outer space is also forbidden (Article IX), so that satellite systems in place cannot be tampered with.

The OST also places international liability on a government for national activities in outer space, whether or not a governmental entity carries them out. Where private parties are actors, the government must provide "authorization and continuing supervision" (Article VI). This is true even if another state, which may be held jointly liable, performs the actual launch (Article VII). These stipulations clearly limit both the ability and the willingness of governments to be aggressively involved in space activities on a private level.

The Agreement on the Rescue of Astronauts, the Return of Astronauts, and the Return of Objects Launched into Outer Space (Return and Rescue Treaty) was passed in 1968.[41] Its relevance for satellite systems is the provision that a fallen space object, or its component parts, will be returned to the state responsible for launching (not necessarily the owner) upon request (Article 5). The treaty also makes it clear that international organizations may be parties to the space treaties (Article 6). The European Space Agency is the only such organization to date.

The Convention on International Liability for Damage Caused by Space Objects (Liability Treaty) has been in force for the United States since 1973.[42] Launching states include, jointly, the state actually performing the launch, if any, the state from whose territory the launch was made, and the state that procured the launching. The Liability Treaty holds the states absolutely liable for any damage caused by a space object on the surface of the earth or to aircraft in flight (Article II). Although discussions during the treaty negotiations centered on the physical effects of collision, the definition of damage in the treaty is quite broad; the only case invoking this treaty involved radiation damage.

The Convention on Registration of Objects Launched into Outer Space (Registration Treaty) was ratified in 1976.[43] It requires nations to register launches with the United Nations, including such information as date and location of launch, orbital data, and general function of the space object (Article IV). The United Nations is also to be notified when objects leave earth orbit. Only one nation may register a given object.

Other International Agreements

Intelsat is an international consortium that is the only party authorized, by the signatories to the Intelsat agreements, to operate worldwide satellite networks for public international communications. Seven nations participated in the initial negotiations to set up Intelsat, which now has 110 members. The organization provides about two-thirds of all international telecommunications.[44] The Soviet Union is not a member.

The preliminary 1964 agreements were not ratified permanently until 1973, after much negotiation but little change. At the outside, the United States had the only satellite network in operation, and Intelsat was organized under the leadership of this country with the explicit goals of a single global commercial telecommunications satellite system, the most efficient and economic facilities possible, and global access to the system.[45] The economic, technical, and political advantages of a single system were clear to all parties, but perpetuation of United States dominance was feared by many. The compromise reached was that ownership of, and investment in, Intelsat was to be determined proportionately by a country's use of the network. That is, a given nation member is assessed a use fee covering a pro rata share of the system's costs. It was also decided that rates for usage would also be based on global averages rather than route-specific estimates. These rates are set by Intelsat so as to provide a given rate of return (now 14%) to its signatories. Thus Intelsat is a cooperative, at its most benign interpretation. When usage is equaled with ownership, the basic tension between the status of nations as users, and therefore wanting low rates, and owners, wanting high rates, helps to keep estimates accurate and the entire system honest.

The initial share of the United States in Intelsat was 61 percent; today it is about 21 percent, with a cap on any possible share of 40 percent. Similarly, ComSat was the original manager of Intelsat and today it is the only United States representative. Technically, individual nations are not parties to the operational Intelsat agreements but, rather, their chosen representatives: ComSat and the PTTs. In a side agreement, governments provide for being bound by their representatives' decisions. In the case of ComSat, which is not a pure governmental agency but has a substantial private character, there have been strong differences of opinion between it and the United States.

The first Intelsat satellite, Early Bird, was launched in 1965. Today it has 16 satellites in geosynchronous orbit under its management, with 850 earth stations in 160 different countries.[46] (Nonmember nations may still receive its communications, and Intelsat often sells excess capacity for nations to operate their own domestic networks.) Although 80 percent of Intelsat's traffic is ordinary telephone messages on the public switched network, it is empowered to handle all forms of communications. Intelsat Business Services (IBS) was introduced in 1983 as a high-quality digital service to carry data, telex, facsimile, and teleconferences, as well as ordinary voice service.

The Intelsat system has come under fire in recent years for its slow introduction of new services. Also challenged is its commitment to a single global system. Agreements require even domestic systems be cleared with Intelsat for technical conformity; that connections to the system can be made only through the government representative, rather than directly by the end user; and that alternative international systems can be set up only after a strenuous showing of no economic or technical harm to the Intelsat network. All these requirements are being contested on some level, but are still in effect today.

The International Maritime Satellite Organization, Inmarsat, was formed in 1978 and became operational in 1982. Explicitly modeled on Intelsat, it handles all maritime (nation-to-ship rather than nation-to-nation) satellite communications for its members. The United States originally preferred these activities to be handled by Intelsat; the Soviet

Union, this time a charter member of the organization, and the United Kingdom held out for a separate organization to increase the size of their participation. Voting and fees are determined as in Intelsat, although at present Inmarsat leases capacity on five satellites (two ComSat, two Intelsat, and one European Space Agency), rather than owning any.

Rather than participate directly in Intelsat, the Soviet Union formed Intersputnik, an analagous arrangement, in 1971. Thirteen socialist countries are members, including Cuba; Belgium is the only country that is a member of both organizations. Intersputnik recently approached Intelsat to negotiate interconnection between the two systems. Intersputnik has also offered its services free of charge to Ted Turner's superstation TBS in the United States; no FCC or Intelsat problems would be posed by such a connection, but the Department of State has asked for reciprocal privileges in the Soviet Union as a precondition.

Other International Organizations

The most important international organization for satellite information networks is the International Telecommunications Union (ITU). The ITU began life as the International Telegraph Union in 1865 as a means of coordinating European telegraph systems. In 1932 it became the International Telecommunications Union, and in 1947 the ITU became a specialized agency of the newly formed United Nations.[47]

The ITU allocates frequency and orbital assignments so as to minimize interference for all satellite communications systems for all of its 160 members. It operates the International Frequency Registration Board (IFRB), which maintains the master registry of orbits and frequencies.

The ITU does not have any police powers as such, and the only sanction at its disposal for a nonconforming satellite is not to recognize its existence. In practice, the ITU offers its services as a mediator in resolving complaints of interference and has an excellent record of obtaining compliance through each nation's self-interest in preventing chaos. Its fixes are typically technical rather than political, and the technological advances of the last decade have, so far at least, outstripped demands on the organization. The greatest current pressure on the ITU revolves around the assignment of geostationary orbital slots. Such assignments were the principal subject of the 1985 World Administrative Radio Conference (WARC), a convention held every five years to set general policy for the ITU. The so-called SpaceWARC did not resolve this issue completely, and a follow-up conference is set for 1988.

Intelsat, Intersputnik, the FCC, and even the military services of ITU members strictly honor ITU assignments as the most basic requirement of operations.

The only international organization that itself carries on activities in outer space is the European Space Agency (ESA). It was formed in 1975 as a merger of two largely dormant organizations, the European Launch Development Organization (ELDO) and the European Scientific Research Organization (ESRO). The ESA is composed of thirteen members (including all of the EEC except Luxemburg) and has a special technical memorandum of understanding with Canada. It developed the Spacelab first used in the NASA shuttle in 1983. More important, ESA is the sponsor of the Arianespace commercial launch program, the only operational competitor to the shuttle. The ESA is the only international organization that is a party to the United Nations space treaties.

At the United Nations, the Committee on the Peaceful Uses of Outer Space is an organ of the General Assembly. The fifty-four-member committee has formulated all the existing outer space treaties. Other United Nations agencies with a more indirect connection with satellite networks include the International Civil Aviation Organization (ICAO), the World Metereological Organization (WMO), the United Nations Educational, Scientific, and Cultural Organization (UNESCO), and the World Intellectual Property Organization (WIPO).[49]

CHAPTER 6
COMPETITION

INTRODUCTION

Most usual forms of competition were not readily applicable to satellite information systems, at least until 1984, for three reasons:

1. Access to, and operation of, space systems was highly regulated, as discussed in the previous chapter.
2. The United States directly, or indirectly, carried out nearly all commercial satellite launches.
3. At every level, the principal actors were nations rather than commercial enterprises.

The deregulation wave of the 1980s, the rise of Ariane, and the private management of space activities around the world have led to a dramatic shift in the possible level of competition, although most would-be competitors are still in the rudimentary stage.

ACCESS TO ORBIT
Satellite Manufacture

The limiting threshold conditions of operating a satellite information network are possession of appropriate satellites and the ability to put them into orbit. Table 6.1 lists the principal United States satellite manufacturers, and Table 6.2 lists the major international satellite manufacturers.

Hughes is by far the world's largest manufacturer of communications satellites, having supplied roughly two-thirds of all those in orbit. Hughes has constructed satellites for all three major applications,

Table 6-1. Major United States Satellite Manufacturers

Hughes Aircraft Company-Space and Communications Group, Commercial
 Systems Division
RCA Astro-Electronics/General Electric, Space Systems Division/General
 Electric, Astro-Space Division
Martin Marietta Corporation
Ford Aerospace and Communications Division
Ball Aerospace-Systems Division
TRW, Space and Technology Group

Adapted from Phillips Publishing Inc., *The 1986 Satellite Directory*, 1986, 193; and from
Reginald Turnill, *Jane's Spaceflight Directory* (London: Jane's Publishing Co., 1984, 300.
Support services, not included here, are those largely unregulated, interstitial activities
that are growing much faster than the "visible" portion of the satellite industry and in fact
are flourishing in this time of comparative retrenchment. For a list and description of
these businesses, see *The 1986 Satellite Directory*, 167–336.

Table 6-2. Major International Satellite Manufacturers

British Aerospace (Britain)
Aerospatiale (France)
Matra (France)
Dornier (Federal Republic of Germany)
Erno (Federal Republic of Germany)
Messerschmitt-Bolkow-Blohm (Federal Republic of Germany)
Aeritalia (Italy)
Japan Communications Satellites (Japan)
Japan Space Communications (Japan)

Adapted from Phillips Publishing Inc., *The 1986 Satellite Directory*, 1986, 319–36; and from
Reginald Turnill, *Jane's Spaceflight Directory* (London: Jane's Publishing Co., 1984), 299–
300.

including Intelsat satellites, as for one-to-one information transfer; a
variety of cable television networks, including its own, as for one-to-
many information transfer; and GOES weather and LANDSAT remote
sensing satellites. In addition, Hughes is part of the joint venture Eosat,
intended to carry on private LANDSAT services.

The other party to Eosat with Hughes is RCA, which runs a distant but not negligible second in volume of satellite manufacture. They
have been aggressive competitors as well, with Hughes successfully
challenging a 1984 RCA-NASA communications satellite joint venture
on the grounds of inappropriate public subsidies.[1] It is no coincidence
that RCA also operates its own networks in one-to-one and one-to-
many systems as well. In the merger mania of 1986, RCA was acquired
by General Electric, and the former's Astro-Electronics Division merged
with the latter's Space Systems Division to become the new Astro-Space
Division of General Electric.[2]

Martin Marietta and Ford have both proposed communications satellites systems[3] to capitalize on their manufacturing capability, but these plans are on indefinite hold. While Martin Marietta's experience is largely in scientific and military applications, Ford has had considerable work in remote sensing and telephony, including helping to set up Arabsat. In July 1986 Ford Aerospace announced that financial pressures necessitated curtailment of its activities; in December 1986 AT&T purchased Ford Aerospace Satellite Services Corporation, the Ford subsidiary intended to run the proposed network and with the FCC licenses to do so.[4] As of late 1987 AT&T had not purchased any part of Ford's manufacturing capability.

Ball has considerable experience in remote sensing and other scientific applications and plans to offer a remote sensing satellite for private use. TRW produces the Tracking and Data Relay Satellites (TDRS) used for military communications and control, among many other systems, and appears to be the only satellite manufacturer not involved in a commercial network.

British Aerospace is the largest manufacturer of communications satellites outside the United States and Soviet Union; its satellites are used for the European Community Satellite Network (ECS) and its maritime analog (MARECS).[5] Taken as a whole, the French space program, with its Ariane operations, is the largest outside the two super powers.[6] With some experience in communications satellites, the French manufacturers' greatest strength is in remote sensing applications, as is that of the Germans.

The two Japanese entities are recently formed consortium. Japan Communications Satellites is a venture of Hughes, Itoh, and Mitsui[7]; Japan Space Communications, of Mitsubishi and Ford.[8]

Launch Services

Only two launch services have placed commercial satellites into orbit: the American Space Transportation Service (STS, more usually known as the space shuttle), operated by NASA, and the French Ariane, operated by Arianespace. A few relevant comparisons of the two are given in Table 6.3.

The shuttle was designed to be able to handle military and scientific missions as well as commercial payloads; these missions required a manned low-earth orbiter. The Ariane, designed from the start for satellite launches, is a more familiar three-stage booster rocket, the latest

Table 6-3. Existing Commercial Launch Services

	Space Shuttle	Ariane-4
Type	Partially reusable manned orbiter	Expendable launch vehicle
Payload	65,000 lb[1]	9,260 lb[2]
Approximate cost per satellite	$25 million[3]	$35 million[4]

[1]From Edward Finch and Amanda Moore, *Astrobusiness* (New York: Praeger, 1985), 131.
[2]From Chris Bullock, "Ariane 4 and Its Competitors," *Interavia* 5 (1986):551.
[3]In 1982 dollars. From Craig Covault, "Economic Competition," *Commercial Space* (Fall 1985):18–21.
[4]From *Space Commerce Bulletin*, 23 May 1986, 8.

and most powerful version of which is designated Ariane-4. The space shuttle has a vastly larger cargo bay that can accommodate payloads such as the Spacelab, for carrying on scientific research and even manufacturing in space, but this additional power does not translate directly into the ability to place more satellites into orbit. The shuttle can reach only low earth orbit, less than 200 miles high. For a satellite to reach geostationary orbit—23,000 miles up—it must be "launched" a second time from the shuttle with the equivalent of the satellite's own third-stage engine. This engineering constraint means that the shuttle has a practical limit of four satellites to be placed into geostationary orbit. This practical limit is further constrained by an economic one. Even before the disasters of 1986, it was judged financially imprudent for insurance reasons to carry more than three satellites on a flight, and two would be a much more common number. By comparison, Ariane-4 is capable of carrying two satellites directly into geostationary orbit.

Estimating the cost of a launch is much more problematic. Both launch services have been heavily subsidized by their national governments, and have traded charge and countercharge over the level of these subsidies. At present, prices approximately cover costs. Decisions made on shuttle pricing in late 1985 changed the price of an entire shuttle flight to $71 million (up from $38 million) until 1988 and $74 million thereafter, all figures in 1982 dollars.[9] The figure in Table 6.3 thus assumes a three-satellite flight. Ariane prices have also doubled in the recent past and are expected to climb at least 10 percent per year, as well as remaining at the mercy of a devalued dollar.

It is impossible to be more precise about these prices because there is effectively no opportunity to launch with either service. The shuttle will not fly again until at least mid-1988. More than half of the commercial flights already contracted for the period 1986–88 will not be accommodated in a new schedule reaching out to 1995; this decision

Table 6-4. Proposed Private Commercial Launch Services

Vehicle	Operator
Titan	Martin Marietta
Atlas Centaur	General Dynamics
Delta	McDonnell Douglas
Delta	Transpace Carriers
Conestoga	Space Services
Liberty	Pacific American Launch Systems
Industrial Launch Vehicle	American Rocket Company

is part of an explicit policy not to launch commercial payloads.[10] An Ariane loss in May 1986 pushed back its program at least one year, at a time when it was already fully booked through 1989.[11] As of September 1987 no firm date had been set for the next Ariane launch (which would be the first of the Ariane-4 vehicle).

This increase in nominal prices, coupled with unavailability at any price, has brought great pressure to bear on the development of the fledgling ELV industry to meet commercial demands. Table 6.4 lists the possible entrants into this field.

The Titan, Atlas Centaur, and Delta delivery systems all exist; have launched satellites in the past, are manufactured and operated by major aerospace contractors, require (relatively) minor modifications to carry large commercial payloads, and hope to receive support from the Air Force's plans to fund and develop alternative launch vehicles. The Titan is the most likely winner of this competition. The Titan was the launch vehicle for the shuttle itself and its most versatile competitor,[12] it has the first commercial ELV contract, for a Federal Express launch,[13] and has had enquiries from six other companies for twenty one launches.[14] General Dynamics has also had discussions with a number of clients but has no firm commitments. A long-standing controversy between McDonnell Douglas, the manufacturer of the Delta rocket, and Transpace Carriers, a private firm created specifically to commercialize the Delta, over receiving commercialization rights and support from NASA was resolved in favor of McDonnell Douglas late in 1986; Transpace Carriers both threatened lawsuits against NASA and initiated direct negotiations with the manufacturer. McDonnell Douglas responded to the decision by reopening its Delta assembly plant.[15]

Space Services is the only entity in the world to have made a wholly private launch (neither government vehicles nor facilities) with the suborbital flight of its Conestoga I in September 1982.[16] NASA has allowed Space Services to plan to use one of its facilities for private launches in late 1988, including one contracted with the Celestis Group, a venture of Florida morticians to place cremated remains into space.[17] Pacific and American Rocket are considerably farther off in their plans. Pacific contemplates a tourist version of a shuttle flight to be offered along with low-earth-orbit launch services; American Rocket is also a low-earth-orbit delivery vehicle.[18]

In addition to these private-sector ventures in providing commercial launch services, several proposals are capitalizing on national space programs (Table 6.5).

Even before the launch disasters of 1986, the Soviet Union had provided governmental launches for other communist bloc nations and India; in 1983 it also offered to put an Inmarsat satellite into orbit, but was turned down.[19] Citing the success of its workhorse vehicle, the Proton—ninety of ninety-seve.ı launches successful since 1970—in 1986 the Soviet Union–formed Gavkosmos, a civilian space agency offering launch services to the world at deeply discounted prices, roughly about $24 million per satellite into geostationary orbit.[20] India has already scheduled a launch, and a contract seems likely with the Iran PTT; discussions are also under way with Intelsat, Inmarsat, and Eutelsat.[21] The United States response to the Soviet venture is that any United States satellite is a piece of sensitive defense technology which cannot be exported to the Soviet Union.[22] Russia has countered with promises of payload integrity and continuous owner supervision.

The People's Republic of China, however, is not on the list of nations denied United States technology. Although China's space experience is much more modest than that of Russia, the Great Wall Industrial Corporation, a commercialization agency of the Ministry of Aeronautics, has been much more successful. Great Wall has contracts signed with three American companies: Pan Am Pacific, Western Union, and Teresat.[23] (Teresat is in the odd position of having a contract but no satellite; the refurbished Westar bird that it intended to buy was purchased, at the last minute, by Pan Am.) In addition to inquiries from fourteen other companies in nine nations, the People's Republic also

Table 6-5. Proposed National Commercial Launch Services

Vehicle	Operator
SL-12,13 (Proton)	Soviet Union
Long March 1,2,3	People's Republic of China
H-I,II; N-I,II	Japan
ASLV, PSLV	India
Hermes	European Space Agency

Adapted from Robert Brodsky, Malcolm Wolfe, and Ian Pryke, "Foreign Launch Competition Growing," *Aerospace America* 24 (July 1986):36–39.

has a launch commitment from Sweden. Launch prices have not been confirmed, but reports range from a claimed $4 million for the Swedish launch, as a loss leader, to about $20 million; China is also providing insurance at about 10 percent rates, compared to 20 to 30 percent in the industry.[24]

Launches are expected to begin in early 1988. The other potential entrants are much farther downstream. Japan's once aggressive launch program has faltered due to both technological and political pressures, and may retreat to handling only domestic traffic.[25] India's augmented satellite launch vehicle (ASLV) and polar satellite launch vehicle (PSLV) development program may be able to place a satellite in geostationary orbit in the 1990s, but is relying on Soviet launch services until then. Hermes is the European Space Agency's version of the shuttle and is not anticipated to be available until 1995 at the earliest; the Soviet Union may have a shuttle by that time, while the West Germans, British, and Japanese probably lag farther behind.[26]

ONE-TO-ONE INFORMATION TRANSFER

Existing Systems

Satellite systems for one-to-one information transfer (telephony) in the United States are operated by a handful of companies (Table 6.6), nearly all of which are involved in offering broadcasting services as well.

All the operators of domestic one-to-one systems are lessors of transponder capacity to at least some degree, usually for commercial applications such as data transmission, video-conferences, and private networks. The few largest common carriers have also integrated their

Table 6-6. Existing Domestic One-to-One Systems

Operator (Parent Company)	Transponders	Type
Skynet (AT&T)	72	Common carrier/lessor
Spacenet (GTE)	80	Common carrier/lessor
American Satellite (Contel)	24	Common carrier/lessor
Satellite Business Systems (MCI)	50	Common carrier/lessor
ComSat	48	Lessor
RCA Americom (RCA)	128	Lessor
Galaxy (Hughes)	72	Lessor
Westar (Western Union)	96	Lessor
Alascom	24	Lessor

Adapted from Phillips Publishing Inc., *The 1986 Satellite Directory*, 1986, 3–84. Domestic networks of countries other than the United States are not included.

satellite facilities into their terrestrial public switched network. AT&T has the largest portion of its network dedicated to telephony, but is also the single biggest competitor in satellite transmissions by way of its terrestrial fiberoptic network. In December 1986 AT&T's purchase of Ford Aerospace Satellite Services to obtain the latter's satellite licences (and possibly to enter into satellite manufacturing) countered suggestions that AT&T was emphasizing terrestrial development at the expense of orbital.[27] GTE broke ground as the first commercial launch carried by Ariane in May 1984, and has the most ambitious expansion plans of any carried. Contel's single satellite anticipated benefits from a merger with ComSat in late 1986, but the merger was called off by the FCC.[28]

ComSat, a carrier's carrier with respect to access to international services, is a lessor of its excess capacity domestically. Its fortunes had been sagging for many years, not least because of its involvement in Satellite Business Systems, an unsuccessful joint venture with Aetna and IBM. ComSat eventually dropped out, then Aetna; finally, IBM sold the enterprise to MCI in March 1986.[29] The purchase gives MCI an entry into satellite operations.

The RCA, Hughes, and Western Union systems, although very large, are mostly committed to broadcasting uses, although they do lease transponders for one-to-one transfers and help to set up private networks. Alascom is a small Alaskan venture providing service within Alaska and linking the state to facilities in the contiguous forty-eight states. Table 6.7 lists the major existing international systems.

The Intelsat system carries roughly two-thirds of global transoceanic telecommunications on its sixteen-satellite system. By definition, competing systems cannot be operated by Intelsat members without its permission, so this list of entities is necessarily quite short. Inmarsat, Eutelsat, and Arabsat have been the three international systems approved by Intelsat, along with an Inmarsat precursor, Palapa (a domestic but transoceanic Indonesian system), and some smaller United States-Canada and United States-Bermuda ventures.[30]

Table 6-7. Existing International One-to-One Systems

Operator	Type
Intelsat	Common carrier/lessor
Inmarsat	Common carrier
Eutelsat	Common carrier/lessor
Arabsat	Common carrier
Intersputnik	Common carrier

Adapted from Reginald Turnill, *Jane's Spaceflight Directory* (London: Jane's Publishing Co., 1984), 209–36; and from Phillips Publishing Inc., *The 1986 Satellite Directory*, 1986, 91–103.

Inmarsat grew out of an early United States system, Marisat, originally operated by ComSat. Its system provides maritime communications for ships at sea and aircraft over international waters on an old Marisat satellite, two Marecs satellites from the European Space Agency, and leased capacity from Intelsat. The Eutelsat system operates three satellites as part of a regional system for Western Europe, but its expansion plans seem to be running afoul of Intelsat's own goals in that area. Arabsat launched two satellites to provide telecommunications services to twenty-two nations and organizations of the Arab League, with some concern over technology transfers from Western manufacturers. Intersputnik is a similar regional network operating independently of, but analogous to, Intelsat (of which the Soviet Union is not a member, though it does belong to Inmarsat); fourteen communist bloc nations are served by two Soviet satellites.

Proposed Systems

Proposed domestic systems are listed in Table 6.8. Given the level of competition and the hard times already existing in the industry, only one of the proposed systems—Martin Marietta—is a traditional telephone system; as noted earlier, motivation probably comes from manufacturing expertise. All the other proposals are in response to the 1985 FCC initiative to allow development of mobile satellite systems.[31] Fi-

Table 6-8. Proposed Domestic One-to-One Systems

Operator	Transponders	Type
Martin Marietta	16	Common carrier/lessor
Global Land Mobile Satellite	12	Mobile telephony
Globesat	1	Mobile telephony
Hughes Communications Satellite Services	36	Mobile telephony
MCAA American Satellite Service	10	Mobile telephony
McCaw Space Technologies	24	Mobile telephony
Mobile Satellite Corporation	NA	Mobile telephony
Mobile Satellite Services	6	Mobile telephony
North American Mobile Satellite	5	Mobile telephony
Omninet	30	Mobile telephony
Satellite Mobile Telephone	NA	Mobile telephony
Skylink	15	Mobile telephony
Wisner & Becker/Transit Communications	1	Mobile telephony

Adapted from Phillips Publishing Inc., *The 1986 Satellite Directory*, 1986, 85–90; and from Richard Anglin, "Mobile Satellites: The New Business in Space," *International Space Business Review* (June/July 1985), 6–15.

nally, a Federal Express proposal to create a satellite system to support its facsimile Zap Mail service is on indefinite hold after the company's withdrawal from that offering.[32]

The mobile system efforts range from that of Hughes, sponsored by the enormous parent company, to that of Globesat, an experiment by the University of Utah. The systems also differ greatly in the number of regions into which a territory is divided. If a satellite emits a single transmission beam, there is great switching efficiency at the cost of re-quiring a unique frequency for each user. With many beams, frequen-cies can be reused from region to region at the cost of switching complexity. For the continental United States (CONUS), for example, the proposed number of beam areas range from one (Globesat, Hughes, Mobile Satellite Corporation, Mobile Satellite Service, and Transit) up to thirty (Omninet) and thirty-seven (Satellite Mobile Telephone).

Despite the rigorous Intelsat restrictions on setting up alternative satellite systems, current United States policy has been to encourage their formation.[33] Table 6.9 lists the proposed international systems re-sponding to this initiative.

In 1983 Orion Satellite was the first to file an application with the FCC to sell or lease transponder capacity on a transatlantic satellite network. International Satellite, RCA Americom and Cygnus followed quickly with similar proposals, although the first two intended to act as common carriers rather than vendors with respect to excess capacity. In 1984 Pan American Satellite (PanAmSat) proposed to offer domestic satellite services to a number of Latin American countries as well as United States broadcasts to the same area. In 1985 Financial Satellite (FinanSat) also presented a varying application for both Atlantic and Pacific coverage for large financial institutions. All applications were ul-timately approved by the FCC, although Orion, Cygnus, and FinanSat were required to make technical amendments to their original propos-als.[34] Still pending are 1986 applications from Columbia Communica-tions for a transpacific system, and by McCaw for a similar Indian/Pacific Ocean network.[35]

Intelsat's resistance to these ventures was as strenuous as FCC approval had been straightforward, but in December 1986 the first steps were taken toward Intelsat approval of the PanAmSat application.[36] Pressure from the United States, the dominantly domestic character of the system, and the absence of a direct challenge to established Intelsat transpacific and transatlantic networks, all played their part in this de-cision. Still under discussion is an attempt by PanAmSat to make use of Cygnus frequency assignments approved by the FCC, but Cygnus is un-likely to be able to make use of them. A direct transfer was ruled out by the FCC, but PanAmSat is now attempting an outright purchase of Cygnus.[37]

Table 6-9. Proposed International One-to-One Systems

Operator	Type
Orion Satellite	Lessor/vendor
International Satellite	Common carrier/lessor
RCA Americom	Common carrier/lessor
Cygnus Satellite	Lessor/vendor
Pan American Satellite	Common carrier/lessor
Financial Satellite	Lessor/vendor
Columbia Communications	Lessor
McCaw Space Technologies	Lessor

Adapted from Richard Colino, "A Chronicle of Policy and Procedure: The Formulation of the Reagan Administration Policy on International Satellite Telecommunications," *Journal of Space Law* 13 (1985):110–12, 119–21, 127–28, and 145–46.

ONE-TO-MANY INFORMATION TRANSFER

Existing Systems

Table 6.10 lists the major users of the active United States systems. In this lengthy litany, two observations can be made. First, no broadcaster operates its own satellite system, nor are there any current plans for one to do so. (RCA does offer service for NBC, to be sure, but also for its competitor, ABC, and, in turn, much of NBC's programming is carried on AT&T's Comstar.) The investment and expertise required seem too great. Second, the major operators have, to some extent, divided up users on the basis of past experience. The RCA network naturally carries most satellite radio transmissions, while AT&T has maintained relationships with all the national networks. Notwithstanding this fact, most users seem reluctant to deal with only one system. For large users such as the national television networks and the premium cable channels, this is probably a form of self-insurance, maximizing the probability of their signal getting through. For many small users, the diversity of systems more likely indicates special opportunities afforded for less competitive slots.

Broadcasting on existing international one-to-one systems is much more limited. Intelsat provides the United States domestic networks, WTBS, the Armed Forces Satellite Network, the Movie Channel, and Cable Network News (CNN),[38] while Eutelsat offers domestic cable services.[39]

Table 6-10. Major Users of Active United States Systems

Satellite (Operator)	Type of Use	User (Parent Company)
Comstar (AT&T)	Cable TV	Country Music Television
		ESPN
		Oak/Telstar
	National TV	ABC
		CBS
		NBC
	Video-conferences	Picturephone
Galaxy (Hughes)	Cable TV	Cable Health Network (Viacom)
		HBO
		Satellite News Channel (Group W)
		Spanish International Network
		Spotlight (Times Mirror)
		Turner Broadcasting
Satcom (RCA Americom)	Cable TV	Armed Forces Satellite Network
		ARTS
		ASCN
		Bravo
		Cable Health Network
		CBN
		Cinemax
		CNN
		C-SPAN
		Daytime
		Don King Sports and Entertainment Network
		ESPN
		Eternal Word
		Financial News Network
		Galavision
		Home Box Office
		Home Sports Entertainment
		Home Theater Network (Group W)
		The Information Channel
		MTV
		The Movie Channel
		National Broadcasting Network
		National Christian Network
		National Jewish Television
		Nickelodeon
		People's Satellite Network
		People That Love

Table 6-10. (continued)

Satellite (Operator)	Type of Use	User (Parent Company)
		The Playboy Channel
		SPN
		Spanish International Network
		Spotlight (Times Mirror)
		Trinity Broadcasting Network
		USA
		The Weather Channel
	National TV	ABC
		NBC
	Other noncable TV	WGN
		WTBS
	Radio	ABC
		Astro Radio Network
		Blue Suede Radio Network
		Bonneville Beautiful Music
		CNN Radio
		Cable Jazz Network
		Family Radio Network
		Georgia State Radio Network
		Gold Mine Radio Network
		Joy Radio
		Moody Broadcasting
		Music in the Air
		NBC
		Rhythm & Blues
		Rock-a-Robics
		SBN
		Satellite Jazz Network
		Satellite Music Network
		Satellite Radio Network
		Seeburg Lifestyle
		WFMT-FM
	Teletext	Reuters
		Time Video Information Services
	Video-conferences	Biznet
		Hi-Net
Westar (Western Union) cable TV		ARTS
		American Network
		Black Entertainment Network
		Blue Max
		Bonneville Satellite
		Cable News Network
		Catholic Telecommunications Network
		Daytime

Table 6-10. (continued)

Satellite (Operator)	Type of Use	User (Parent Company)
		The Disney Channel
		Hughes Television Network
		Madison Square Garden Cable Network
		Nashville Network
		Robert Wold Communications
		Satellite News Channel (Group W)
		Selec TV
		Spotlight (Times Mirror)
	National TV	ABC
		CBS
	Other noncable TV	PBS
		WOR
		XEM

Adapted from Anthony Easton, *The Satellite TV Handbook* (Indianapolis, Ind.: Howard W. Sams, 1983), 305–27.

Proposed Systems

Domestically, the significant proposals are all for direct-broadcasting satellites (DBS) systems, encouraged by recent FCC initiatives and ITU easing of regulatory restrictions. The DBS systems seem extremely unlikely to be sanctioned internationally, except within an already approved regional network such as Eutelsat.[40] Table 6.11 lists the major current ventures.

Table 6-11. Proposed Domestic Direct Broadcasting Systems

Operator	Transponders
Advanced Communications	16
Cablesat General	NA
Direct Broadcast Satellite	12
Dominion Video Satellite	16
Hughes Communications	32
National Christian Network	12
Satellite Syndicated Systems	6
Satellite Development Trust	NA
Space Communications Services	NA
United States Satellite Broadcasting Company	24

Adapted from Phillips Publishing Inc., *The 1986 Satellite Directory*, 1986, 85–90.

The proposals have two things in common. First, all involve two satellite systems for national coverage. Second, none is close to actual implementation, and, with the exception of Hughes, all of the largest players in the first round of proposals—ComSat, CBS, RCA, and Western Union—have dropped out.

Somewhat closer to launch, after many years of negotiations and setbacks, is a European system, Coronet, to be operated as part of Eutelsat.

MANY-TO-ONE INFORMATION TRANSFER

Existing Systems

Just as with launch services, only two existing earth observation systems have commercial access, one French and one American. Also, as with launch services, the existing American entity does not exist for the moment. Eosat, the RCA-Hughes joint venture intended to be a private governmental LANDSAT program,[41] has repeatedly sought additional governmental funding to carry out this task. These requests were continually denied, and even the originally proposed funding level was the victim of budget cuts. In December 1986 Eosat stopped all work on LANDSAT 6 and 7 and began to lay off its employees. Negotiations are under way to attempt to salvage plans for at least one satellite.[42]

As Eosat's fortunes sank, those of Systeme Probatoire d'Observation de la Terre (SPOT) were on the rise after a slow start in early 1986. Paradoxically, SPOT was able to obtain a contract with the United States government to supplement, and perhaps later to guarantee the continuity of, the LANDSAT archives; after the agreement, SPOT opened two earth stations in Canada to allow faster processing of material for the United States.[43] Even if Eosat is to reenter the game, SPOT, with system in place and contract in hand, is likely to remain far ahead.

Proposed Systems

No other commercial remote sensing systems are being planned, although Japan, Brazil, Canada, and India all have governmental systems under consideration. Japan plans to launch its Earth Resource Satellite (ERS)-1 late in this decade.[44] Brazil, a long-time heavy user of LANDSAT data, is developing a SPOT-like vehicle of its own for deployment in

1989 as part of an all-Brazil space program (Missao Especial Complete Brazileira).[45] Canada's Radarsat, not limited by ground weather conditions, is scheduled for the early 1990s.[46] India has had previous, though unsuccessful, experience with remote sensing satellites and intends to develop a second-generation system, Insat-2, to be built and launched by itself.[47]

Although no private domestic remote sensing ventures are under consideration, several network proposals qualify as many-to-one information transfer in the area of radio determination. The most advanced of these is Geostar, more famous for its founder, the space activist Gerard O'Neill, than for its service: a pocket-sized transmitter to a satellite relay system. RCA is manufacturing the relay, and the project is expected to be launched as part of an upcoming GTE-Ariane package.[48] Geostar reports the position of, and a short message from, the transmitter; originally planned as an emergency/rescue device, it is now thought to have better applications as a property locator.

Two other firms—McCaw and Omninet[49]—have included such a system as part of their mobile telephone proposals, but are much farther away from implementation.

Part 3
Satellite Information Systems: Issues for Decision Makers

CHAPTER 7
Issues for Users

GENERAL ISSUES

Certain issues face all users of satellite information systems, regardless of the type. These general issues are raised by broad trends often affecting the telecommunications industry as a whole, including political, economic, and technological trends.

Political Trends

The almost total deregulation in the United States of almost all aspects of satellite communications over the past decade has increased options for the user in at least four ways. First, deregulation of operators is effectively deregulation of uses of satellite systems. This has led to the availability of many specialized services for the user. Most dramatic among these has been the almost overnight creation of a secondary market in resale and leasing of satellite capacity. Second, removal of barriers to entry in the domestic arena, and the desire of the Reagan Administration to do so internationally as well, has increased the number of competitors among whom a user must choose and has created a (temporary) glut in raw satellite capacity, especially for one-to-many broadcasting. Third, these same factors blur the distinction between user and operator by making private networks so easily implemented. It is a very small jump from the backyard dish to the rooftop antenna. Last, there is almost no regulation of reception: the most basic "use" of all.

All these opportunities present confusion to the potential user, confusion fueled by the lack of any historical performance record. (Consider the number of competitors, let alone services, mentioned in chapter 6, which did not exist five years ago.) Indefinite launch delays and the certainty of an industry shakeout should prompt a conservative user to deal with those operators large and well established enough to have a reasonable chance of surviving the next three years.

This same deregulatory movement is also taking place in most Western nations. At the same time, many developing nations have called for a New World Information Order (NWIO) that would stringently regulate both communications reception and transmission as well as guarantee access to satellite network technology (among other areas) to all comers.[1] Chief NWIO concerns are in controlling propaganda entering, and information leaving, a country. These concerns have retarded, or have attempted to retard, national participation in many countries where, paradoxically, more extensive use of satellites for information distribution would aid development. Satellite technology has no need to recognize borders and national sovereignty, and local governments may feel that they have lost control.

What has also been lost, however, is the fear of satellite systems' role in a "big brother" or "eye in the sky" extension of governmental authority. The explosion of uses and the miniaturization brought about by technological advances have caused pervasive decentralization, rather than centralization, of access and power. Once any private use is permitted, the genie is let out of the bottle and all uses can become private. These same decentralizing and pluralistic effects also seem to preclude dreams of a "global village" of integrated networks. Even though worldwide broadcasting is now a reality, the most appropriate image for satellite communications is, instead, the tower of Babel.

Economic Trends

Contrary to expectations, the push to deregulation has slowed and, in some areas, completely halted a long-standing trend in service rates. While the need to meet competition exerts some restraining pressure on rates, the large number of new competitors in the field has led to enormous duplication of costs. The growth of the secondary market in resale and leasing of satellite services, although it makes new opportunities available to many users, has also imposed a new layer of costs. Finally, although start-up investment requirements continue to decline, although not as sharply as in the 1970s, they are still quite substantial in absolute terms for new ventures.

These same high entry costs, coupled with launch uncertainties, have made the cost improvements in satellite information systems almost invisible to the user. If he is dealing with a new operator, he must support a large up-front investment. If he is, instead, dealing with an established network with the resources to commit to a new venture without the need for immediate return, then the incremental value of the new venture will be small compared to the costs of its older, embedded base of systems.

For the operator, it almost never makes economic sense to abandon existing systems in favor of more efficient ones; additions to systems typically come about in response to increases in demand (and the outright failure of old satellites). With the intense competition for users, it is hard for any operator to command a market share large enough to sustain expansion on these grounds alone. Large business users have been able to counter these forces somewhat by becoming their own private networks. In any event, many users feel that the availability of services formerly not purchasable at any price outweighs the fact that current prices may not fully reflect cost reductions.

In many ways the satellite industry is entering the same slump in which the computing industry, long thought to be unassailable, has wallowed for the past three years.[2] The dropping of the Department of Justice antitrust suit against IBM, the surge in competition spurred by overnight legends of Silicon Valley successes, and the resultant glut in the ultimate decentralized market—the personal computer—as well as in the chip-manufacturing industry all parallel the current state of the satellite business. There is no immediate happy resolution in sight.

Technological Trends

The principal contributions of technological advances to satellite systems have been the increase in discrimination of reception and the attendant reduction in power transmission requirements. Simultaneously, the ability to discern weak signals and the ability to focus a signal narrowly for the downlink have increased, so that a relatively small (1 meter!) dish can clearly receive signals only obtainable from a dish ten times its size a few years ago. Thus, power demands for transmission are decreased, or, by holding power constant, more sophisticated uses at higher frequencies are possible. Advances in telemetry, which help to keep dish antennas properly oriented, have also magnified the effective capabilities of small earth stations. Finally, lower power requirements decrease possible interference, not only with other satellite signals but also with ground communications.[3]

The decreased size requirements for earth stations have also brought down the costs of these stations. Dish installation is now a thriving cottage industry associated with hardware shops, appliance stores, and even mail-order kits.[4] The cost for a do-it-yourself receive-only station is now well under $1,000 and is still descending.

All these benefits are largely due to the advances, made in the 1970s, in power and complexity of microelectronic chips, which have made possible "smart" satellites, dishes, and coffee machines. Although the advantages of existing technology are far from exhausted in the satellite field, the rate of improvement is clearly decelerating. Recent price reductions will not be matched in the next few years, as the early 1980s lag in chip research finally catches up with actual implementation.

ONE-TO-ONE INFORMATION TRANSFER

Privacy

The privacy of one-to-one information transfer is very strong in fact (due to the volume of such communications) but quite weak in practice for at least four reasons. First, the legal protection afforded such communication is uncertain. FCC regulations[5] do prohibit the unauthorized interception and use of any licensed transmission. These regulations, however, are firmly set in the context of wiretapping; that is, of some physical connection to the individual communication. Amendments of 1985 address interception of cable television signals with respect to actual hookup to the cable, not interception over the air.[6] Where communications are carried on telephone wires, interception is well defined and illegal. Where the transmission process is microwave radio signals between earth stations, the integrity of the signals seems to be in legal limbo, but at least the interceptor must be in line of sight with the transmission. Interception of satellite signals is completely unprotected. It is also uncertain whether a user of certain modalities—a cordless telephone, mobile telephone—or, for our purposes here, a lessee of a satellite transponder is in fact the transmittor of the communication in the eyes of the FCC. If so, he is certainly an unlicensed transmittor and not entitled to any protection at all.

Second, it has been ruled that interception of satellite telephony signals from *outside* the United States does not violate any domestic law.[7] Although this decision was made in the context of upholding an interception by law-enforcement authorities, it seems to have deadly implications for international financial transactions.

Third, for satellite communications there is no possible way of knowing that signals have been intercepted. There is also no way of impeding interception for real-time interactive communications; however, one-way communications could be speeded up and transmitted in rapid burst form to help avoid detectability. The only effective means of defeating interception is by coding or scrambling the signal at the site of the actual transmission, at great cost.

Fourth, there is no recourse, and rightly so, against the operator of a network for intercepted signals; that is, no operator of a satellite system acts as a guarantor of the privacy of its communications.

In short, a user of satellite communications services cannot know whether his signal has been intercepted and can probably do nothing either legally or technically if he does know. Only the volume and usual unintelligibility of most communications are working to protect him. In special cases, such as national security communications and sensitive computer-to-computer linkages, especially electronic funds transfer, scrambling of signals over private networks is routine to provide a measure of protection. At the same time, these steps make the communications easier to target.

It should be stressed that the vast majority of interceptions are unintentional, but as users, operators, and levels of service involving retransmission of the signal proliferate, such interception will increase.

Reliability

The flip side of the interception problem is interference: that is, unwanted interceptions by the user. Historically, this has been only a minor issue for one-to-one information transfer, since not only the intended communication but the connection itself was one-to-one, a unique physical link from sender to receiver. Such interference as does occur with traditional twisted-wire pair communications is most commonly due to the electromagnetic fields generated by even the weak power in telephone lines. Satellite systems and the current telecommunications environment contribute to increased interference in several ways.

First, since telephony by satellite is carried on by radio waves rather than a physical connection, some spillage from competing signals is inevitable, as is the influence of local atmospheric conditions. Reception technology, however, is at the point where this type of interference is represented at most by occasional static. What satellite communications are not well protected from is the "hash" at ground level created by short-range radio transmissions from cordless and mobile telephones, citizens-band radio, and even video cameras and remote-control gadgetry. These unpredictable interferences, not under the control or observations of the satellite system, can overwhelm discriminatory power of reception.

Second, the separation of local from long-distance telephone service, brought about by the AT&T divestiture in 1984, has compromised coordination among the satellite systems operated by AT&T and GTE, the portions of networks leased by other carriers, and the local exchange companies still involved in the final physical links of the communications process. While the lack of integration itself adds to interference possibilities in a general way, the real danger is in the number of uncontrolled handoffs from one system to the next. Delivery processes are being splintered and layered without the authority or the opportunity for control of overall signal integrity.

Third, the divestiture itself places the local exchange companies at a disadvantage by simultaneously precluding the local companies from involvement in satellite systems and promoting the advantages of satellite systems operators (i.e., AT&T and GTE) in bypassing local delivery altogether.[8] Dealing directly with the system operator will aid signal reliability for those customers able to engage in their own reception and processing. Even such a customer will still be dependent upon the local carrier for most of its message traffic, and resultant pressures will cause local service crunches, affecting signal reliability for all customers.

Linkages to Other Services

Integrating conventional telephone services with more specialized one-to-one information transfers (e.g., electronic mail, facsimile and data transmissions, video-conferences, and the like) as well as now routine business operators such as word processing, reproduction, inventory control, and, above all, office computing, has long been a gleam in the eye of communications writers. One catchphrase for this wholesale integration is "office of the future." Unfortunately, that "future" seems to be the same as that of the World's Fairs and popular magazines of the past fifty years: always a perennial generation away, moving sidewalks and all.

The office of the future is as far off today as it was in the 1960s, when the phrase was first coined. Satellite information systems, coupled with advances in microcircuitry, are one reason this will remain so. This is because the power of programmable chips, and the reach satellite systems give to specific products and services, always seem to outweigh the incremental net benefits of integration due to integration's cost, inflexibility, and novelty.

Integration is inherently costly because it does not profit directly from the raw power or scope now available to telecommunications. An ideally efficient and adaptable power source might be sufficient in theory simultaneously to run a furnace, a refrigerator, and three cars for a household; the much harder part would be managing the connections and the controls. Just so, a single satellite transponder may be more than adequate for a business's worldwide voice, data, and video communications without making it obvious or practical to organize them all at once. Progress in integration of systems by definition always lags behind that made in any system component. Furthermore, until a perfect world arrives, the integrated systems pioneer will always have to cope with all the unintegrated systems that have not caught up to him yet. This also detracts from the ultimate cost benefits of integration.

The need for each system component to interact smoothly with all others requires system interfaces to be somewhat inflexible compared to the opportunities afforded by unlinked services. The first experience of most users with fully integrated communications and computing services is that exactly the combination of most interest to them is unavailable. With on-satellite processing power always improving but still always at a premium, satellite systems again are not the easy answer and, indeed, are an obstacle to total integration.

Finally, the innovativeness of integration is a critical impediment to its implementation. This is true not just in the trivial sense of novelty associated with any new activity, but for two additional reasons. First, there is no one in charge; there is no entity or operator to whom to appeal who can impose order from above. On the contrary, local telephone service is legally divorced from the long-distance communications where satellite systems have their worth; the largest satellite operator's involvement in many specialized services is regulated to the extent that these services must be kept separate from traditional communications. Before the AT&T divestiture, integration support was at least theoretically available to the individual user. Today it is usually more than he can manage to get to a predivestiture level of coordination, much less improve on it. Most users must cope with a retrogression in systems integration instead.

Even if these regulatory obstacles did not exist, integration of communications systems would require considerable innovation in the minds and habits of users. Integration is a real break with the past, and familiar behavior is changed very slowly—much more easily for one new service at a time than for all at once. Reluctant users typically expend more effort in trying to circumvent the system than in trying to maximize its usefulness alone. Conversion time and training costs are often enough to stop integration proposals in their tracks.

This litany of negative arguments has two possible exceptions. Integrated linkages with other one-to-one information transfer services seem to be modestly effective where the communication is one-way or integration is accomplished *ex nihilo*. Examples of the first case are on-line (typically bibliographic) data systems and remote control/alarm services. Here, near total control of the communication is in the hands of the active user; at the very least, the passive partner's behavior is completely predictable. Accordingly, the user is always dealing with known types of signals with a very small repetory. Examples of the second case include so-called smart or telecommunications-ready buildings, or even whole industrial complexes, such as Teleport, on Staten Island in New York City.[9] In these instances, the project developer is the effective integrator, creating an integrated set of services as a fait accompli for the user, without any previous mix of modalities requiring tedious conversion. Often such projects are unhindered by continuing commitments and large enough to bypass connection with the local exchange company almost completely, and thereby bypass other barriers to integration as well. One drawback is that the user is tightly constrained by the quality and flexibility of the developer's system.

Dedicated Networks

Again and again, issues confronting the user have centered on the decentralization and pluralism, if not outright chaos, in the operation of satellite systems. This diversity was the dominant factor in the political, economic, and technological trends discussed above; the chief source of difficulties in achieving private, reliable, and integrated communications; and it is best represented by the rise of dedicated networks. By dedicated networks is meant private (satellite) systems that restrict both the types of services offered and the users with access to the system. Unlike common carriers, which restrict neither, and the so-called specialized common carriers, which restrict only services, these networks usually revolve around the needs of a particular user such as Federal Express's Zap Mail service, and may be so narrowly defined as to blur the distinction between one-to-one and one-to-many communications such as a satellite service for a grocery chain that provides Muzak and inventory control information.[10] The networks, in the long run, are wasteful of capacity, but every new system makes it that less likely that the trend can be readily reversed. The only effective damper on dedicated networks for the near future is the difficulty of getting them into orbit.

Transborder Data Flows

Transborder data flows are "the transmission of computer data across international frontiers by means of computer-to-computer communications."[11] Where this communication takes place on physical telephone connections it is at least nominally at the mercy of the local telecommunications operator, who in turn, in nearly all countries other than the United States, is a governmental agency. Where the communication is by direct connection to a satellite network, there is no opportunity for local control; even monitoring is difficult, given the uninterpretability of data compared to voice communications.

Many developing countries have proposed, but none has implemented, restraints on such communications, which are almost always carried on by Western businesses operating locally. The usual concern is that information necessary to regulate domestic business activities is being sent, and kept, outside the country; less typically, that sensitive national information, including data about citizens, is being sent abroad where it will not be protected. The most concrete and realistic of the proposals is one raised by several Latin American nations: that any data sent out of the country must also be maintained and identified in an in-country data base. A particular sore point is the ability of foreign banks, with their superior communications, to arbitrage a country's own currency. Nevertheless, such a proposal, if broadly implemented, would only reverse the priority of fears, with a user having to defend the proprietary nature of his own data.

At one extreme, the problem is viewed as a purely technical one: regulate the technology that makes the communication possible and the communication is regulated as well. Many business users will feel, however, that it is only this advanced communications technology that makes it possible for them to operate internationally; thus, a ban on rapid information transfer would be a ban on business altogether. Some countries see restriction of transborder data flow not as an end in itself but as an instrument of national industrial policy. Such restrictions impede financial and communications businesses with little domestic investment much more than they do large manufacturing operations that contribute to the local economy.

At the other extreme, it is argued that information is a good and should be controlled as any export. Such a proposal has not made much headway, due to horrendous problems of definition and enforcement, but has received support from an unlikely source. The regulations defining and controlling the European Common Market call for free exchange of all "goods" among member nations. In a broadcasting dispute, such transmissions were held to be goods in this sense, and not subject to governmental interference from the *sending* country.

Consideration of this issue is scheduled for the next round of the General Agreement on Trades and Tariffs talks, under the heading of "intangible goods." The likelihood of any new strictures in this forum is very low, but the very discussion will lend credibility to those countries wishing to control transborder data flow unilaterally.

ONE-TO-MANY INFORMATION TRANSFER

Control of Content

Although networks apply extensive self-censorship, there is almost no overt control of programming content for broadcasting in the United States. On rare occasions network affiliates have refused to carry particular programs, including episodes of prime-time series, the content of which was judged offensive to community values. More frequent have been refusals to carry public-service programming, in the form of news or documentaries, which the local stations simply find too boring. A compromise is usually struck between the local affiliate and the network for a package of popular shows and special-event programming together with the public service broadcasts.

Much more effective in controlling content is the outcome of a community's decision to award a cable television franchise. If the cable operator and the municipality cannot agree on terms, users are effectively debarred from any cable programming. Often part of the negotiation process is the allocation of several cable channels to community access (typical local news or educational programs), again controlling content to some degree. Until late 1985, an FCC regulation held that cable operators must also broadcast all normally available local stations. Known as the "must carry" rule, it was overturned in litigation on first amendment grounds, holding that cable operators should be free to determine their own mix of programming. The industry is in the process of developing a voluntary version of the must carry rule on a sliding scale, where small operators (fewer than twenty channels) need carry no local stations and large operators must carry all.[12] By imposing this level of self-regulation, the cable operators hope to forestall congressional legislation on the topic and to avoid pressure for the same from the broadcast industry.

Control of content is also achieved indirectly by local restrictions on access by satellite dishes. Today such restriction can only be based on health, safety, and esthetic grounds, despite pressure from cable operators for more general exclusion.[13]

Even nominal access to a cable operator's system is no guarantee that the user has access to all the programming he desires. Most premium cable services (that is, movies and sporting events) are owned and produced by the cable operator himself. Some combinations of programming will therefore never be available, due to the competitive relationship of the carriers. Each year bills have been routinely introduced in congress to require cable operators to divest their owned services and in effect become common carriers. Just as routinely, these bills are turned down.

The advent of DBS would obviate most of these forms of control, bypassing both municipal intervention and the cable operator, although perhaps simply substituting another programmer in its place. Although many firms are at least discussing DBS, technical and cost problems still loom large, and the revolution is not near.

Unauthorized Reception

Unauthorized reception of pay television services is illegal in most states through a theft of services argument when a physical connection is made to the cable itself. For many years, however, the question of whether reception of signals "in the clear" by a backyard dish constituted signal piracy was in legal limbo. Its recent resolution in favor of the user (i.e., the decision that freely broadcast signals are freely available) prompted the obvious response by cable operators. Scrambling of signals had been long decried by the operators as too costly to be justified; now it is merely necessary. By the end of 1986 most premium services (HBO, Showtime, ESPN, The Disney Channel, the Playboy Channel, and the Movie Channel, among others) were sent scrambled and decoded only in the user's home.[14] The case for coding pay services is obvious; most operators are scrambling all their basic service programming as well. Unfortunately for the user, one of the national networks (CBS) will also scramble its signals, and the other two are considering the decision. Even some local stations (such as New York City's WOR-TV) will no longer be sent in the clear, but will require decoding.

Users' legal suits against operators have been fruitless. Many claim that dish reception is the only means of access available to them if they are in an isolated area beyond any local broadcasting signal, and believe that they should not be penalized for this circumstance. Others claim willingness to cooperate with the cable operators but that dish decoders are unavailable due to the enormous demand created by the scrambling decision. Neither argument has met with a warm response.

That cable system operators should wish to increase their potential market by scrambling signals is understandable, but for a national network to do so seems counterproductive since, in large part, its revenues are determined by the size of its audience. The networks advance two reasons.[15] First, their direct transmissions carry material not intended for the public user, including several views of the same action or setting, and blank spaces for local announcements and commercials. Second, they wish to honor agreements with sports teams on the local nonavailability of sporting events (blackouts). The real reason has to do with the exclusivity rule.

The FCC regulation of cable programming was originally based on two principles: the must carry rule (discussed above, and now no longer in effect) and the exclusivity rule (still on the books although proposed for deletion by the FCC). When a local station purchases or rents a piece of syndicated programming, that station typically receives the exclusive right to show the program in a certain market for a certain period of time. If the cable operator carries the same program, normally by rebroadcasting a distant signal from outside the local market, the local station's advertising revenues are harmed by the competition.[16] On the other hand, the cable operator is perfectly indifferent to the number of people watching any program, and may even claim that he has increased the value of the rebroadcast signal.[17] The FCC's exclusivity rule decides this question in favor of the local station, although it has not been strictly enforced due to studies showing that cable's impact on local station revenues is small for this particular type of competition.[18]

In this context, the decision of networks and local stations to scramble their signals is an attempt to seize control of rebroadcasting from both the FCC and the cable operators. If fully implemented, many broadcasting uses now taken for granted will become unauthorized (and unavailable) receptions.

Unwanted Reception

Simple interference is the most common case of unwanted reception for users of one-to-many information transfer. The domestic and international regulatory system of orbit and frequency allocation is generally effective in preventing interference by competing broadcasting signals, although problems exist for some small European countries where broadcasting footprints are larger than the size of the nation. Much more of a problem is competition from ground-level microwave transmission of telephone conversations.[19] Terrestrial relays of these communications (in addition to radar and other microwave signals) cause most interference with satellite reception, which is much more resistant to problems from surrounding obstacles than typical local broadcasts. Microwave transmission interference can usually be rectified by small realignments of the dish and by use of special filters.

Most unwanted reception problems are social rather than technical and so do not have a correspondingly easy solution. Even where it does not directly conflict with local signals, uninvited broadcasting over national borders is often seen as a cultural imposition at best, and as propaganda at worst. These reservations are always on the part of governments (France and Latin America) rather than their populations. These same concerns are also the critical stumbling block in negotiations for a DBS treaty at the United Nations.

Private Networks

The history of one-to-many information transfer has been one of boom or bust with respect to channel capacity. The current and near-term situation is clearly a boom period, with some systems able to carry greater numbers of different broadcasts than exist in the entire world today. Instead of requiring stringent licensing procedures for new broadcasters, the industry is now furiously but unsuccessfully seeking new sources of programming. Furthermore, just as in the case of one-to-one systems, dozens of private, idiosyncratic networks are springing up rather than a few monolithic and all-inclusive ones. Decentralization and duplication rule the day here as well, but the anticipated diversity of material has failed to materialize.

MANY-TO-ONE INFORMATION TRANSFER

Few special issues arise for the user of many-to-one information transfer since these services are still in their infancy, but two areas deserve at least brief mention.

Effects of Private Ownership

The effects of private ownership of the United States LANDSAT program and its first competition in the offerings of French SPOT services will be to provide the typical user with both more and less information than before. Eosat, the domestic joint venture of RCA and Hughes Aircraft, which is now operating the United States remote sensing network, will be providing data less rapidly and at a higher cost than LANDSAT; it may drop certain services or portions of the satellite network altogether if anticipated revenues do not appear; and it forbids duplication of the data that it provides clients.[20] At the same time, Eosat intends radically to increase the sophistication of its sensing equipment in the near future, to work aggressively in promoting enhanced data services, and to accept requests for specialized runs and processing.[21] Eosat is hedging its commitments to the general public with strong hopes for developing a two-tier market also aimed at large business users. SPOT is directly aimed at corporate clients and presently offers more sensitive and frequent mappings.

Casual use of the remote sensing network will sharply decrease due to the cost and delay factors; priority service is available, but at a substantial fee. Profiting most from these new arrangements will be users with specialized needs and the ability to perform a portion of the necessary processing themselves.

Confusion of Public and Private Uses

Outside the United States, almost all users of the LANDSAT system (prior to its private status) were other governments. Information was given to them on an at-cost basis, as well as free assistance in constructing receiving earth stations in an attempt to defuse sentiments that remote sensing represented an intrusion on the sensed country's rights. While these relationships are grandfathered into the agreement with Eosat, they probably will not apply to new data and services to be provided.

In any event, negotiations between Eosat and foreign governments, with the United States as an awkward partial partner, will be difficult for many reasons, not the least of which is that most governments are using LANDSAT data for what would be considered private uses in America, principally oil and mineral prospecting. Complicating matters still further is the increasing use of these networks by private parties for an ostensibly national purpose: espionage. Private corporations, especially the aerospace industry, scholars, and news media are directly privy to intelligence matters, given the more refined resolution power of SPOT and the marketing initiatives of Eosat, and fears have even been voiced of use of the network by terrorists.[22] Given the confusion in both directions of public and private uses of the system, the very quality of the data produced may paradoxically lead to increased restriction on its distribution.

Chapter 8
Issues for Operators
GENERAL ISSUES

Operators of satellite information systems are typically users of those systems and in that role are subject to the same concerns expressed in the previous chapters. All operators face other issues as well, regardless of the type of system involved. These general issues are raised by broad political, economic, and technological trends often affecting the tele-communications industry as a whole. The focal point of almost all of these issues can be summarized in one word: access.

Political Trends

Access to existing satellite systems, and the ability to create new ones, have been enhanced by recent political attitudes. In the United States, broad presidential policy directives have intended to advance the commercialization of space. These include statements of public support for the private launch industry to promote ELVs as first, an alternative to and later, as a substitute for the space shuttle in putting satellites into orbit; the decision to replace the Challenger with a new fourth orbiter for the shuttle fleet; the private ownership of the LANDSAT (remote sensing) program; the designation of the Department of Transportation as lead agency for coordinating federal licensing of space ventures; and the wholesale deregulation of the telecommunications industry, intended to increase competition both domestically and internationally.

These deregulation efforts have been largely successful and have even seen similar moves in other countries.[1] The progress actually made on other fronts is much less positive. The ELV industry, while appreciative of generally stated support, desires more tangible help in the form of favorable tax treatment, increased access to federal facilities and information, greater intellectual property protection for their development efforts, and government orders for their services as well as outright subsidies; neither the administration nor congress has been forthcoming on these issues.[2] The replacement shuttle[3] and the remote sensing[4] initiatives have been only partially successful in receiving funding. No new licensing regulations have been issued by the Department of Transportation to guide new ventures.

These gaps in performance leave the current administration in the position of neither actively promoting nor actively retarding an independent launch program. Accordingly, if not for the pressure the Challenger disaster placed on the development of alternative orbit delivery systems, operators of satellite systems would be relatively unaffected by these decisions. On 15 August 1986, however, President Reagan determined that the launching of private satellites "can be done cheaper and better by the private sector," and ordered NASA to phase out its commercial launches.[5] Of forty-four flights contracted for commercial payloads, only one flight in each of 1988 and 1989 would be allowed, with a total of fifteen available by the end of 1992. Part of the administration's intention is that owners will unilaterally decide to move many satellites scheduled for the shuttle to private ELVs, but it is unclear whether any alternative private vehicle will be available in the near future. At the same time, NASA is battling with the administration, the Air Force, and the private sector for expanded authority to launch its own ELVs.[6] The legality of denying existing contracts and the method of choosing eligible flights are also uncertain.[7]

The loss of commercial access to the shuttle is in large measure due to a diversion of launch resources for military needs, which may even spill over into ELVs. For the near term, that is, until at least 1988, when the shuttle is currently expected to resume service, pressure to replace short-lived reconnaissance satellites is mounting.[8] For the longer term, military missions were anticipated (even before the Challenger disaster) to occupy about 75 percent of the shuttle's schedule, up from a historical figure of about 33 percent,[9] with roughly half this amount dedicated to Strategic Defense Initiative flights.

During this pause in United States activities, other nations (France, Japan, and the Soviet Union[10]) are pursuing plans to develop their own shuttlelike systems, with China and India[11] also joining in the competition for national ELV services. The United States government has strongly resisted any suggestions of an American Arianespace to launch commercial satellites for any party, but is still required by the Outer Space Treaty to provide "supervision and regulation" of its national activities in space.

In summary, the political trends affecting all operators of satellite systems tend to be rather negative for the short term (three to five years), with the possibility of a sharp turnaround once the availability of access catches up with the opportunities of deregulation.

Economic Trends

Three economic issues retard development of new satellite information systems for the near term:

- A transponder capacity glut,
- The unavailability of launch insurance, and
- The difficulty of raising capital for space ventures generally.

Availability of transponder capacity on satellite systems tends to be highly cyclical, given the long lead time needed to put a satellite in orbit, the rapid competitive response to technological advances and regulatory retreats, and the impossibility of adding incrementally to existing capacity—a fraction of a satellite cannot be launched. As a result of these factors, demand for transponders exceeded supply in 1979 but, by the end of 1983 only half of existing satellite capacity was in use.[12]

Since that time, underuse probably increased, at least until the spate of launch disasters in early 1986. Transponder demand has been estimated to be anywhere between 50 percent[13] and 80 percent[14] of capacity at that time. This overage will provide a modest buffer for the impacts of launch delays, but probably for not much longer than late 1988. Given that neither the shuttle nor the ELV industry will be in business by that time, and that Ariane is fully booked for the forseeable future, the much discussed transponder glut is about to become a transponder shortage. Should an operator wish to capitalize on this midterm opportunity, he would find himself unable to secure any launch insurance for his satellite, just at a time when it seems more necessary than ever.

The normal actuarial experience with a new type of venture is for frequent losses to occur in the early years, with a stable, low rate of loss gradually achieved over time. For satellite launches, the experience has been exactly the reverse, confounding (and essentially bankrupting) the space insurance industry.

Prior to 1984 the insurance industry had paid out roughly $200 million in five isolated instances for satellite failures.[15] In February 1984 two satellites on a single shuttle flight did not reach orbit, with an insurance loss of nearly another $200 million from that isolated incident.

The insurers, NASA, and the satellite manufacturers teamed up and successfully retrieved both malfunctioning satellites a few months later, but the hope of eventual resale of the refurbished birds has yet to materialize.[16]

Another loss in 1984 and a sudden rash of five losses by September 1985 created an additional $400 million debit for the insurance industry, with a cumulative loss ratio of 2:1; that is, by Fall 1985 twice as much had been paid out in insurance losses as had been received in premiums. The launch failures of early 1986 might have exacerbated this problem slightly—only one insured satellite was involved—but for the fact that the space insurance industry had already disappeared.

Insurance rates had been as low as 5 percent and had climbed to 20 percent of a satellite's value just before the last of the 1985 disasters, and to 30 percent just afterward. Rates this high scared away many operators, and at the same time there was almost no available pool of insurance funds for an operator even willing to pay this amount. In November 1985 a $75 million RCA satellite was launched uninsured.[17]

Into this void other parties have offered insurance of a sort, while the nominal insurers are attempting to implement modifications to their procedures that would allow them to go back into business. Arianespace will offer a second flight for a premium of 11 percent of the cost of the launch, provided a failure has to do with the launch itself.[18] Before the Challenger disaster in February 1986 NASA's policy was to offer a free second flight for any failure due to a shuttle fault, and a second flight at half price if a satellite failed for any reason within ninety days of launch.[19] Manufacturers of both satellites (Hughes) and launch vehicles (General Dynamics, Martin Marietta) have discussed offering insurance for failures due to their components at a premium of about 16 percent, but these proposals have not moved past the talking stage. Many parties outside the insurance industry have called on the government to act as an insurer of last resort, perhaps temporarily, charging a rate higher than the highest commercially available rate.[20] Budget-conscious legislators have not found this proposal attractive.

Possible technical fixes on the insurer side include the following:

- Foregoing insuring of numerous payloads and limiting the length of commitment before launch[21];
- Partial government indemnification[22];
- More aggressive salvage and product liability litigation procedures[23];
- Industrywide pooling procedures modeled on private nuclear power insurance[24]; and
- Technical simulation of launch failures.[25]

A final source of economic pressure is in raising funds for a new space venture. Except for the handful of companies that can easily risk hundreds of millions of their own capital, raising private investment is made difficult by[26]:

- The enormous capital requirements involved;
- The longer than usual payback period for a financial return (eight to ten years, as opposed to three to five); and
- The political and technological risks involved (as in the discussion immediately preceding and following this subsection).

These constraints almost certainly doom any of the proposals for private purchase or manufacture of an additional shuttle orbiter over and above the built-in economic inefficiencies of the manned shuttle as currently designed.[27]

Technological Trends

The first half of 1986 saw an unprecedented number of launch failures which appear to be technologically based. These failures have caused far more than the short-term delays associated with the immediate losses; they also constitute the motivation to develop alternative, more reliable launch systems (with an attendant increase in delay of availability).

On 28 January 1986 the space shuttle Challenger exploded shortly after liftoff, killing the seven persons on board and destroying the payload of a government satellite.[28] Subsequent investigations attributed the failure to a leak in a joint of the shuttle booster rockets, which allowed hot gases to escape. Redesign of the shuttle launch vehicle to correct this error is expected to postpone shuttle launches until 1988, at a minimum. A Titan missile, believed to be carrying a photo reconnaissance satellite, exploded at Vandenburg Air Force Base on 18 April. The destroyed satellite was probably the last KH-11 satellite available to the military, the long-lived generation of all-purpose keyhole spy satellites. Its intended successor, the KH-12, can only be launched by the shuttle and was designed to be refurbished by shuttle flights as well,[29] so that the United States surveillance capacity was dealt a double blow by this failure.

One week later a Nike-Orion rocket blew up at White Sands, New Mexico, on a minor military flight. Eight days after the Nike loss, however, a Delta rocket, launched by NASA on 3 May 1986 to place a new GOES weather satellite into orbit, lost power and was destroyed shortly after blastoff.

The shuttle, Titan, and Delta rockets are the only United States delivery systems capable of putting satellites into geosynchronous orbit or of launching large payloads generally: The near simultaneous loss of flights on all three vehicles effectively shut down both commercial and governmental launches.

The United States was not alone in these troubles. Less than four weeks after the Delta loss, on 29 May, France's Arianespace was forced to destroy an errant rocket carrying an Intelsat satellite.[30] At that point, every current means of putting a commercial satellite into orbit was made suspect. Although an official cause was determined only for the Challenger disaster, sabotage was ruled out for all of these incidents. Instead, quality control and design flaws were deemed responsible. A single failure might have been ignored by governments and industry; the actual plague called all procedures into question while causing havoc in launch schedules. Among other effects, it caused operators to lose their insurance coverage for future flights, if and when resumed, with the prospect of receiving any replacement coverage nonexistent.

The troubles were still not over. On 23 August 1986 an Aries rocket carrying an x-ray telescope was destroyed when it went off course shortly after launch, due to the failure of a small electrical component.[31] This was the last failure of the year, with successful September launches of a weather satellite and "Star Wars" experiments.[32]

The need for alternative launch vehicles set off a three-way race among NASA, the Air Force, and the fledgling ELV industry. NASA expressed interest in operating a mixed fleet of shuttles and unmanned rockets, if only to preserve customers for the shuttle.[33] Recently proposed amendments to NASA's budgets would make this shift a requirement for continuing business, but congressional motivation is assistance to the ELV manufacturers from whom NASA will be forced to purchase launch services. But such purchases will place NASA in direct competition with the Air Force's own program for missile procurement.[34] At the same time, the Air Force has delayed plans for the operation of its own shuttle until at least 1992, and is actively funding development of three new launch vehicles for heavy, medium, and light payloads.[35] To complete the circle of ironies, the MLV (medium launch vehicle) is a new version of the Titan rocket, earlier supported by NASA, which will likely become the new shuttle launcher and its chief Air Force competitor.[36] The immediate benefactors of these policies are most likely the major aerospace manufacturers, at the expense of the would-be private launch services companies. The minimum five-year development period for new rockets, however, represents a vacuum that other fixes, including political pressures and competition from other modalities (especially new ocean cables), will rush to fill.[37]

ONE-TO-ONE INFORMATION TRANSFER

The Challenge to Intelsat: Will the International Network Exist?

Since its creation in 1964, Intelsat (International Telecommunications Satellite Organization[38]) has carried on almost all international satellite telephony outside the Soviet bloc. In the past few years, a series of challenges has arisen questioning Intelsat's position as the only, or even the paramount, network for providing such services.[39]

The agreements bind their signatories (all nations) to obtain Intelsat permission for a member country (or any entity within the member country) to use non-Intelsat space systems for international public telecommunications service. Permission will be granted where Intelsat finds technical compatibility with its own network and no "significant economic harm" to either existing or planned services.[40] Similar permission is needed to operate even domestic systems.[41] All domestic requests have been approved straightforwardly and, until 1983, the eight requests for alternative international networks were approved after modest negotiation. These included new regional networks (Eutelsat, Palapa, and Arabsat, covering Western Europe, Indonesia, and the Arabian peninsula, respectively); other global networks (Marisat, the United States maritime system; Inmarsat, its international equivalent; the use by Algeria of Intelsat facilities to connect with Intersputnik, the Soviet bloc counterpart of Intelsat); and two special use systems between neighboring countries (television broadcasts to Bermuda from United States domestic systems, and a United States-Canada transborder data link). Because these systems replaced existing terrestrial networks or carried small amounts of traffic to remote areas, they were not viewed as economic threats to Intelsat.[42]

Emboldened by the transborder data decision, in March 1983 Orion Satellite Corporation, a new venture, initiated a proposal to the FCC for a new private satellite network between the United States and Europe to carry corporate traffic. The Orion filing was followed quickly by similar proposals from International Satellite, Inc. (ISI), Cygnus Satellite, RCA Americom, Pan American Satellite Corporation, Systematics General Corporation, and Financial Satellite Corporation. These proposals claimed to offer services not provided in Intelsat, to serve areas not served by Intelsat, and/or to allow direct connection with Intelsat (rather than the seeming legislative requirement that connections be made through ComSat). Citing foreign policy issues, the FCC postponed consideration pending White House review.[43]

The potential applicants had three obstacles to overcome. The first was the ComSat enabling legislation, dating from 1962, which established that entity as the only United States operator of international satellite systems. An escape clause in the legislation states that an alternative system may be appropriate where, after a presidential determination, it "is required to meet unique governmental needs, or is otherwise required in the national interest."[44] While this exception was traditionally thought to address the need for military communications and reconnaissance systems, in November 1984 President Reagan announced that separate international satellite systems were required in the national interest, largely due to the benefits of competition. The presidential determination specifically excluded connection with the public switched telecommunications networks, thereby avoiding direct competition with the bulk of Intelsat traffic, and was made contingent upon Intelsat acceptance.[45] The FCC quickly extended the ban on interconnection to all levels of resellers and users of the applicant's services.

Although it is the nominal United States representative to Intelsat, ComSat took issue with administration policy, which has the effect of challenging much of its revenues, and continues to battle the issue before the FCC and with the Departments of Commerce and State. Neverthless, the presidential determination cleared the path to meet the second obstacle, Intelsat authorization after a finding of no significant economic harm. Intelsat has taken a very firm stance against the proposals, claiming that they conflict with new or proposed Intelsat services on high traffic routes and are mere cream-skimming incursions.[46] Not only is United States voting power in Intelsat at an all time low, but the American intiative is seen as a challenge to the foundations of the organization, so approval is extremely unlikely.

A third, hidden obstacle to new systems is that new services must receive the permission of the receiving country as well. This has been received only for the case of Pan American Satellite and Peru, and Intelsat has now approved that application as well.[47] There is nonetheless fear on Intelsat's part that, in return for favorable rates, other nation members may make special deals with the applicants and aid their cause.[48] While the financial costs of the long delay are thought effectively to eliminate all the applicants except RCA, there is also Intelsat concern that AT&T is waiting in the wings should a schism occur.[49] In the meanwhile, special pieces of legislation have been introduced to challenge (or eliminate altogether) ComSat's standing as gateway to Intelsat.[50]

The delay in launches has provided an unexpected cooling off period for controversy that threatened at its inception to undermine Intelsat, one of the two mainstays of international satellite networks. (The other, the ITU, is also under attack, as will be discussed.) That threat may have eased, but only at the cost of forestalling less ambitious alternatives for many years.

Divestiture's Aftermath: Does the Domestic Network Exist?

The effects of the AT&T breakup have been exacerbated by satellite systems so that a single, integrated, domestic telecommunications network no longer exists for one-to-one information transfer. The operations, technology, economics, and even definition of such transfers have been changed to create several (and sometimes incomplete) networks, united only by their commitment to satellites.

Operationally, the AT&T divestiture decree resulted not only in a split between provision of local services (remaining a monopoly) and long-distance communications (open to competition), but also a financial incentive for avoiding transfers between the two in order to escape network access charges (and network-averaged rates). The way out for customers was seen to be creation of private, dedicated networks, bypassing all or part of the pre-1984 system. A few large customers may construct their entire networks from scratch; most will participate in some private network pieced together by an alternative common carrier, or mix public and private facilities to create a virtually private network.[51] All such networks trade off range of services and calling areas for price and control benefits, and in most cases the new network and/ or the bypass operation itself are achieved through satellites.[52]

Technologically, the advantages of a fully digital network are overwhelming.[53] They are not fully realized, however, until an entire network—or a separable piece of it—becomes digital. Rather than upgrading of facilities piecemeal, use of new digital satellite systems is the more attractive course for AT&T and an almost necessary step for its competitors: instant digital network.[54] The largest opportunity at stake in this context is the conversion of the private United States government network to digital capability.[55]

Economically, the distance-insensitive costs of satellite communications (in fact, the almost negligible marginal cost) have unleashed a league of competitors for long-distance services. All-satellite systems are much less costly, only if filled to capacity; resultant price wars have begun to winnow the field of proliferating systems.[56]

Finally, even the definition of telecommunications has raised issues that, when combined with idiosyncracies of satellite systems, have further splintered the notion of a unitary domestic network. The boundary between communications and computing has always been a thin one, and threatens to be eliminated altogether by the capacity and digital capabilities of new satellite systems. At the same time, satellite networks' only real drawback is the time lag associated with interactive voice communications, a drawback that does not exist for machine-to-machine communications. But AT&T has also been forced by regulation to separate operations and rates for basic (simple switching of voice messages) from enhanced (processing with some value added, typically of data communications) services, and has been barred altogether from direct sale of services that might be construed as electronic mail.[57] The company has also been banned from publishing ventures as well, but it is an open question as to whether this prohibition encompasses on-line data base inquiry systems.[58]

Network Integrity: Can Operators Protect Themselves?

Ensuring the privacy of one-to-one communications is a concern of telecommunications network operators as well as their users. Doing so for satellite systems is made more difficult by technological gaps, legal gaps, and national security pressures.

Technologically, the most secure two-way communication would take place on a shielded cable joining both parties to the conversation. As the number of parties using the same channel increases, and as more of the transmission is broadcast by microwaves rather than taking place over a physical connection, the possibilities of interception increase. Satellite networks for one-to-one communications carry an enormous volume of conversations and are within line of sight of almost any interested party, so that preventing interception of messages is essentially impossible technologically. The only feasible alternative is to encode the messages transmitted so that they are unintelligible to any eavesdropper. Doing so increases costs and degrades accuracy, although digital systems can minimize these drawbacks.

Legal pitfalls also are associated with signal interception, both in the definition of what constitutes a crime and in the possibilities for law enforcement. Wiretapping laws are well established but only protect interception on the basis of a physical connection.[59] Issues originally raised by cases involving cordless telephones are now seen to apply to satellite communications: they receive no legal protection at all.[60] For domestic communications it is at least conceivable that such a law could exist. Intercepting—or inserting—messages outside of national boundaries altogether (i.e., in space itself or over the oceans) would be much more problematic to outlaw, in addition to its almost total undetectability. This is a special fear surrounding international funds transfers. Again, encryption (here, primarily by users) is the only protection, with no legal recourse against the culprit. In fact, even between parties, the international legal confusion is often so high that losses related to bad transfers are typically just written off.[61]

Finally, national security pressures complicate both the technological and legal gaps, with governments wanting to straddle the issues. The ease with which satellite signals are intercepted has led to government-prompted advances in network integrity, especially for military systems.[62] At the same time, to maximize control of transborder data flow as well as reconnaissance and law-enforcement opportunities, it is in government's interests for such interception to be routine, at least by the right parties. When a satellite system is used to bypass a domestic network, typically government-controlled outside the United States, regulation of multinational business is also involved.[63]

ONE-TO-MANY INFORMATION TRANSFER

Geostationary Orbit: Who Controls the Sky?

The geostationary orbit—that orbit over the earth's equator where a satellite's speed matches that of the earth's rotation and the satellite appears motionless with respect to a point on the earth's surface—allows a single satellite to broadcast to one-third the planet, and three satellites to set up a global network.

In geostationary orbit (roughly a band a few miles wide but 22,300 miles up) the closer two satellites are placed together, the greater the danger of collision. Long before danger of collision becomes non-negligible, however, the broadcasts of the two satellites will begin to interfere with each other. While such crowding is still in the future, except for satellites intended to cover the continental United States, the number of satellites that can ultimately be placed in geostationary orbit has some upper bound.

Slots in geostationary orbit have been assigned, up to now, on a first-come, first-served basis after approval of orbit and broadcast frequencies by the ITU (International Telecommunications Union). This has prompted concern by many developing countries that they will have no opportunity for an orbit of their own, and that the placement of satellites in geostationary orbit constitutes an appropriation of territory forbidden by the Outer Space Treaty. That treaty forbids national appropriation under any claim of sovereignty, but also precludes any interference with another party's peaceful use of outer space, of which communications satellites are probably the best example.[64]

Ironically enough, the first major challenge to unfettered use of the geostationary orbit was a statement of ownership by other nations. The Bogota Declaration, signed by eight equatorial states in 1976,[65] claimed that geostationary orbit was linked to a particular spot on the earth's surface; that this connection was based exclusively on gravitational phenomenon; and that, therefore, it was not part of outer space at all but, instead, a piece of the nation below. Milder forms of the assertion pressed the argument that geostationary orbit was a "limited natural resource" and part of the "common heritage of mankind," these latter arguments joined in by the nonequatorial developing countries. The scientific basis for the claims was quite spurious, but the political motivations were real and varied. In some cases there was a genuine desire to reserve slots for future national use; in others, to obtain payments from foreign satellite operators in a national orbit; in still others, merely to guarantee some sort of access to use of the orbit; or, finally, to use the declaration as a bargaining chip in the battle for control of programming.

The ITU first directly addressed these concerns in its 1979 World Administrative Conference (WARC) with general rulings that a right to an existing slot did not guarantee any rights in perpetuity and that geostationary orbit was, in fact, "a limited natural resource" that must be used economically and efficiently to allow equitable access to it by all countries.[66] The same conference also scheduled a 1985 WARC as a planning session and a 1988 WARC for implementing a structure for managing the geostationary orbit.

The catchwords for the 1985 WARC were "allotment," a priori assignment of slots, favored by the developing countries, and "access," an opportunity to buy into existing or future networks on nondiscriminatory terms, favored by the developed countries. After an extremely bitter conference that left many key issues unresolved, a compromise was reached leaning strongly toward the allotment position.[67] For an expanded set of frequencies, every member nation of the ITU was to receive an orbital slot dedicated for domestic use; international systems, operating on existing frequencies, were to have improved methods for determining access. Agreement could not be reached on an actual allotment scheme or the future improvements, and determination of both was forestalled till the 1988 WARC. This failure to reach agreement and the acrimony that surrounded the session have alarmed many observers of the industry, who see continued technical (as opposed to regulatory) fixes for expanding usability of the geostationary orbit as the only means of avoiding a civil war within the ITU.[68]

Direct Broadcast Satellites: Who Controls Transmission?

Direct-broadcast satellites (DBS) take advantage of enhancement in power and quality of transmission of television signals to allow reception by a small rooftop antenna two or three feet in diameter. The DBS have been hailed as a means of bringing broadcasts to remote regions at minimal costs.[69] The most highminded systems planned are typically modeled on an experiment in the Canadian Northern Territory in the late 1970s, bringing news, educational, and cultural programming, all produced by the government, to outlying areas.[70] Other proposals concentrate on straight entertainment programming on a pay-television basis. For those areas too distant or thickly settled to support cable systems, DBS could be a godsend, especially given the recent decision to scramble broadcasts by the national networks.

At the same time DBS have been denounced as a threat to national sovereignty, as outright propaganda, and, by the Soviet Union, as a use of the mass media for purposes of ideological competition. What is feared is if the capacity were to exist to beam broadcasts (from a foreign nation) directly into a citizen's home, bypassing any involvement with the terrestrial network under his government's control. Even where television networks are government entities, as in most countries, governments have chosen to control distribution rather than program content directly.[71] This enormous tension between freedom of information, seen as an abstract value, and the immediate threat to national sovereignty delayed even discussion of DBS for years, until the ITU, in 1983, first gave its authorization to solely domestic DBS systems.[72]

This decision came shortly after the United Nations General Assembly passed a resolution in December 1982 stating that all nations have the right to veto any incoming television broadcasts by satellites from abroad.[73] While the resolution acknowledged "the right of everyone to receive and impart information and ideas as enshrined in the relevant United Nations documents," this right was in effect dominated by "the principle of nonintervention." While the resolution is not binding, it also called for an international treaty on the subject. Such a treaty is being negotiated by the United National Committee on the Peaceful Uses of Outer Space (COPUOS), with essentially the same terms under consideration.

The political uncertainties of DBS have combined with the economic risks to put off implementation for at least the near future. Although there had been many first-round applicants when the FCC first invited proposals in 1982, most dropped out upon a requirement to show financial viability, and nearly all the rest when the FCC added a requirement of "due diligence" in pursuing the venture.[74] Effectively eliminating any short-term opportunities, in late 1984 ComSat, besieged by its other problems, broke off from its own DBS venture, taking a significant loss.[75] ComSat had been the first (as well as by far the largest) entity to propose direct broadcasts. Effective control of transmissions remains with national governments.

Signal Ownership: Who Controls Reception?

Operators of satellite broadcasting systems are very sensitive to any suggestion of regulatory control of their programming, but they received a threat from a new direction when the national HBO broadcast was interrupted on 27 April 1986, and viewers east of the Mississippi saw *The Falcon and the Snowman* replaced by this message:[76]

> Good Evening HBO
> From Captain Midnight
> $12.95/Month?
> No Way!
> (Showtime/Movie Channel Beware!)

The HBO operators tried to fight off the interfering signal by increasing the power of their own, without effect, and Captain Midnight's pronouncement ended after five minutes.

The interloper turned himself in a few months later, but his brief appearance on the air prompted disclosures that all of the nation's commercial programming, and a large portion of its military traffic as well, were vulnerable in exactly the same way.[77] Armed with the knowledge of a satellite's exact position and receiving frequencies—matters of public record in required filings to the FCC, the ITU, and the United Nations—a dedicated hacker needs only to give his signal more power than the true source. A more vindictive party might send up enough power to burn out the satellite; a more knowledgeable one might send signals to change the satellite's position. Constant two-way monitoring of the signal with a fallback alternative available is the only solution, unless, of course, the pirate has two transmitters.

Captain Midnight's raid was in response to the cable operators decision to scramble their signals.[78] Scrambling alone, however, is unlikely to be able to combat rebroadcast of satellite signals by other nations domestically. For United States operators, this is a particular problem in Latin America, where satellite footprints inevitably extend well beyond United States borders.[79] Satellite operators charge piracy and signal theft; other governments counter with claims of unwanted, or least uninvited, transmissions. Copyright treaties are weak or inconsistent with respect to such broadcasts. The one treaty explicitly covering the matter—the 1974 Brussels Convention on Distribution of Program-Carrying Signals by Satellite—prohibits any rebroadcast of a signal, though not the unauthorized reception itself.[80] Unfortunately, the United States is not a party to the convention.

The residual formal regulation of program content and reception—domestic copyright and must carry rules[81]—constrain an operator only marginally. A broadcast satellite system operator has few legal or technological safeguards for ensuring that the signal received is the signal he sends, however, or that the parties receiving it are those to whom it was sent.

MANY-TO-ONE INFORMATION TRANSFER

Limits on Observation: Sensed Country Rights

Remote sensing operations as carried on by governments—and until 1986 only the United States LANDSAT program existed—naturally emphasized the nonintrusive benefits that the sensed country received. Meteorology and disaster prediction and relief are obvious cases where the sensed country is anxious to receive data, although some reservations or embarrassment may be associated with publicity given to disasters. Agricultural data seem to fall into a middle ground, typically providing little excitement for either the sensed or sensing country. Detecting and measuring natural resources, however, is resented and feared by many countries, with private remote sensing ventures only aggravating these feelings.

At one extreme, sensed countries believed that the sensing is a violation of their nation sovereignty and an implicit appropriation of their natural resources. Where exploration took place necessarily on the surface, a state could control exploitation fairly easily. No such control is possible for them of discovery by satellite. Furthermore, they claim that since remote sensing exclusively depends on, and is motivated by, these surface measurements, it is not an activity "in" space that is protected as a peaceful use of the Outer Space Treaty but is instead unlawful on its face. Less extreme positions call for prior consent as a precondition for remote sensing, for the ability to control dissemination of information to third parties, or to have priority in obtaining the information originally.

The United States' counter to these arguments has traditionally been its open skies policy, with the belief that space is open to all peaceful uses. The United States has also attempted to defuse the political sensitivity of its remote sensing operations by providing access to data received to all parties on an at-cost basis and by providing foreign nations with assistance in constructing receiving stations for LANDSAT signals.[82]

These initiatives, however, are not reconcilable with commercialized operations where the need to make a profit dictates a higher level of control. Commercial operators also place a premium on distinguishing primary raw data, presumably provided to all parties, from secondary enhanced or value-added data, available only under special conditions and perhaps only on a commission basis.[83] In addition, they cite the impossibility of screening out a particular country's data, even should they wish to. Instead, they are seeking some form of copyright protection for the extra level of processing they provide.

Recently, the Legal Sub-Committee of COPUOS reached agreement on a set of draft principles to govern remote sensing.[84] These principles do not give a sensed state veto power over sensing, but do provide that it have access to both primary and processed data on a nondiscriminating basis and on reasonable cost terms. Sensing states would also be required to consult with sensed states to "enhance mutual benefits." The draft principles do not have the force of law, but are the basis for a treaty being negotiated by COPUOS.

Limits on Observation: Domestic Regulation

As discused in chapter 4, Eosat, the private American successor to LANDSAT, is to operate under narrow constraints to minimize changes in United States policy. In addition to these formal requirements, funding and first amendment problems also loom.

As of November 1986 Eosat is still warring with congress over its initial funding.[85] Congress has been withholding the small part of Eosat's request for appropriations it approved pending reports from the Commerce Department on commercialization. Eosat, in turn, has begun to scale back, or eliminate altogether, plans for its first satellite. Although negotiations have repeatedly approached the brink and then receded, the associated delays continue to increase what seems to be a minimum eighteen-month data gap between the failure of the last LANDSAT satellite and the launch of the first Eosat.

At the same time, the government is concerned that private parties will make uses of a private remote sensing system that either intrude upon governmental functions or constitute an unacceptable level of invasion of privacy. Pictures of an alleged Soviet shuttle base were one of the first sales of the French SPOT service, and were commissioned by a private research company for sale to the news media.[86] This was merely the latest in a series of civilian spying incidents that are likely to increase due to the improved resolution and private control of the next generation of remote sensing satellites.[87]

A secret executive order prohibits planned United States systems from operating below a 10-meter resolution level, and straightforward cost issues prevent private efforts from fully competing with military spy satellites.[88] This order and subsequent Commerce Department regulations appear to conflict directly with first amendment protection of the press.[89] News agencies are certain to stretch the boundaries of allowable behavior in the near future, and it is likely that the news media may operate their own remote sensing system rather than rely on images generated by any other party.

Epilogue
A Brief Look at the Future
INTRODUCTION

Some trends for the future of satellite information systems are clear and natural extensions of recent experience: continued technological enhancements, growth in the number of players, and increased pressure for access and control by developing nonspace countries. The future is also to be shaped by a number of events that have acted as shocks to the system, such as the 1986 string of launching disasters, the wholesale deregulation of the telecommunications industry, and the Strategic Defense Initiative. The impact of these unanticipated events is more than enough to undermine the credibility of any alleged forecast.

What is offered here is a brief personal view of the near-term future (roughly to the year 2000) of the development of satellite information systems. Three alternative futures are sketched, each one a picture of what *might* occur if certain trends and events dominate. None of them in isolation is likely to be the real future, but in combination they come close to spanning the range of possibilities. It is also probable that the real future will possess some of the attributes of each alternative in an ambiguous and evolving world. Some major event, unforeseen here, will almost certainly occur over the next thirteen years, but an attempt has been made to be complete as to the types of effects any event might have.

Finally, two of the three views discussed are relatively pessimistic for the short-term future of satellites. That proportion, just as all the judgments in this chapter, is only a matter of the author's personal opinion.

THREE SCENARIOS

The three "what if" scenarios presented here are loosely constructed around the broad groups of environmental issues described as political, economic, and technological trends in the preceding two chapters. The categories are gross oversimplifications and ignore the fact of relationships among the clusters; for example, a failed economic scenario probably presumes a failed technological one. Nonetheless, these categories represent separate and often extreme forces and values that shape satellite (and all other enterprisory) activities. Accordingly, their separate treatment is hoped to be more revealing than misleading.

The very idea of a satellite information system contains very strong assumptions as to these scenarios. Satellite communications are essentially apolitical, at least to the extent that national borders are completely irrelevant. They are elements of transnational, global networks that at least contain the seeds of ignoring or obviating governments altogether. This form of independence understandably makes nation-states uncomfortable, and anxious to create some form of control. At the same time, this insensitivity to boundaries makes the satellite a privileged weapon and an intelligence-gathering device all too likely to be drawn into the embrace of national security. These sovereign interests form the basis of the political scenario.

Satellite systems also possess nearly unique economic attributes as well. Very expensive to deploy, they are entirely distance insensitive and almost completely volume insensitive; that is, high embedded costs with close to zero marginal costs. Therefore service prices reflect much more of a need to recapture investment rather than continuing expenses. Thus, if launch costs are brought down; *or* longevity prolonged so that embedded costs are more easily recovered; *or* the level of use is high enough for the cost of initial investments to be smeared across a large base, then satellite information systems come close to being free goods. Uncertainty as to launch, life, or use eradicates all these advantages, which is the basis of the economic scenario.

Finally, satellite systems are necessarily high-technology activities whose success is conditioned on ever-accelerating progress. Should this progress be interrupted by political or economic realities, or should a ready use for the progress not exist, its advantages are negated. A positive answer to this problem is the basis of the technological scenario.

A Political Scenario: Militarization, Nationalism, and Retrenchment

In the political scenario, national interests prevail. For the nations with space capability, this means that most new access to orbit is dedicated to military uses. For the nonspace nations, this means guaranteed access and/or control of existing systems. Unchecked, these twin pressures would cause a severe retrenchment in satellite information system development.

Until very recently, the Soviet Union and the United States were the only nations with the technological expertise and surplus capacity to launch commercial payloads. While the Soviet Union has pursued an almost exclusively military satellite program, the United States has, from the very start, relied on a separate civilian agency, NASA, to support commercial activities. This civilian, commercial emphasis has been compromised by the announcement of the Strategic Defense Initiative (SDI), which, by many estimates, could absorb all present and potential launch capacity well into the future; the executive decision to phase out commercial payloads on the shuttle altogether, beginning with the removal of already contracted and scheduled flights; the loss of the shuttle Challenger and other vehicles, with the attendant launch delays; budgetary pressures seeming to preclude the building of a replacement orbiter; policy decisions denying direct support to the ELV industry; military dominance of future shuttle flights; and military uses of the space station.

These needs are real and uninterruptible; launch delays have already compromised the United States reconnaissance satellite network. But if the hiatus in availability of launch vehicles continues, fulfillment of those needs may crowd out commercial flights altogether. For the other nations that possess, or are developing, launch capability, selection of payloads will become increasingly politicized under this scenario, will receive little support from the United States, and will still be inadequate to meet short-term commercial demands.

These pressures will undermine any potential regional efforts, just as the success of Ariane has exacerbated tensions within the European Space Agency. The increasing demands of developing states for access to, or operation of their own, satellite networks will collide head on with decreasing opportunities, as well as current United States policy to withhold, rather than increase, commitments to international efforts. At their most extreme, these forces of militarization and nationalism could effectively undermine the global networks that do exist. At a more realistic level, they are almost certain to bring about a sharp retrenchment in the development of any new or expanded systems.

An Economic Scenario: Commercialization, Competition, and Confusion

Ironically, if the SDI materializes slowly or not at all, there will be a surplus of launch capacity rather than a shortage (provided that most announced activities come into being, another big assumption). Should this occur, satellite networks would temporarily benefit from the short-lived price wars that would follow.

Other likely consequences of free-for-all competition are less favorable for satellite systems, and most are associated with the prospective gap in launch services. The number of proposed systems is so large that, were it possible to place them all into service immediately, the transponder glut would be unsupportable and most new ventures would collapse immediately. As the gap increases, the weaker proposals will fall by the wayside, leaving, in theory, a smaller and more profitable market for the survivors. During this same interregnum, however, active players will be turning to other modalities, especially fiberoptics, to deploy their networks. A fiberoptic transatlantic cable will be placed into service about the same time that commercial flights are expected to come into service, and will sharply cut into an important satellite route as well. Satellite system operators may be kept on the ground economically due to the lack of insurance and to fears of future control, or even appropriation, of their activities by international regulators and/or developing countries.

Finally, the American ELV industry continues to move in fits and starts, depending on the level of anticipated government support. If delays increase, demands will become more selective and volatile, leading to further delays in a vicious cycle.

In this scenario, satellite systems are not militarized but remain largely commercial. Due to competition on the ground and in the skies, however, new development is confused and slow for economic reasons.

A Technological Scenario: Private Ownership, Digitization, and Decentralization

In this scenario it is assumed that governmental needs do not absorb all new satellite development, and that technological improvements overpower any short-term economic negatives. In the complex, real, non-scenario world, it would be quite possible for the technical advances ultimately to result from militarization, rather than to occur in its absence.

Additional transmission power and discrimination available for the next generation of satellites will allow more specialized uses for more direct users without interference, in either broadcasting or regulatory terms. One recent dramatic development in satellite station keeping can lengthen life in geostationary orbit several times over by the simple expedient of incorporating small, predictable movements in the receiving antenna, rather than being forced to spend precious fuel to keep a satellite exactly on track. "Antenna farms"—space platforms with several operators and systems—can accommodate the needs of many activities with the expense of a single launch.

All of these factors—increased power, varied uses, elimination of the middleman, longer life, and sharing of launch and operation expenses—mean lower costs. In turn, lower costs mean the opportunity for higher volume, which drives costs still further down, and so on. This is the flip side of the negative economic scenario, where higher costs drove away users and operators alike.

In addition to these technical fixes, which are largely differences in degree, digital satellite systems represent a difference in kind. The next generation of digital satellite systems will allow direct competition with terrestrial fiberoptic networks; increase transmission speed, security, and discrimination (increasing potential volume of use, as above); and allow new services.

This scenario is predicated on the frankly optimistic assumptions that users exist to take advantage of new systems and services, and that technology will outrun regulation. If these assumptions hold, then the implicit agenda of satellite information systems—transnational, low-cost communications—will be realized. This means wholesale private ownership of satellite systems under the loose administration of existing international cooperatives and regional associations. The final consequence will be decentralization of access to information in ways yet unimagined.

A POLITICAL SCENARIO: MILITARIZATION, NATIONALISM, AND RETRENCHMENT
One-to-One Information Transfer

With no new launch capability at hand, entrenched operations for one-to-one satellite systems can only benefit. New satellite networks are the most effective way of dealing with the leviathan AT&T; if they are impossible or indefinitely postponed, AT&T will remain entrenched as the industry giant and the possibilities for successful competition will be much weakened. The only other party who could stand to profit is GTE, which operates a satellite system larger than AT&T's, although its telephony operations are smaller, and has experience as a privileged customer of Ariane. New services, such as mobile telephony, would also be on indefinite hold. The first trickle-down benefits of a militarization program might be improved security and reliability of communications. Some accelerated growth in fiberoptics networks would also occur.

On the international front, potential American competitors to Intelsat would be engaged in an increasingly bitter struggle over rights to systems that cannot exist simultaneously because of limited access, rather than limiting economics. This starker dimension of the conflict, fueled by nationalist pressures on all sides, could prove to be the undoing of the fragile Intelsat association. With greater certainty, the demands of developing nations with respect to access to geostationary orbit would increase as well, with no opportunity to be satisfied in the near future, leading to a less powerful and more politicized ITU. In summary, there would be a strengthened but more narrow single domestic network, with greatly weakened international networks.

One-to-Many Information Transfer

If we read "the national networks" for "AT&T," this scenario is much the same as for one-to-one information transfer immediately above. The major television networks use the AT&T system predominantly and, so, will not be compromised. Similarly, their competition, the cable industry, requires active development of new systems to mount a significant threat. In this case, however, the cable satellite networks on the Hughes and RCA systems actually anticipated the majors' use of satellites, and they are themselves sufficiently entrenched so that they will lose ground slowly, if at all. Some attention would naturally shift to other local distribution methods.

Internationally, this scenario would suggest more moves toward control of content in every arena. Direct-broadcasting satellites would almost certainly never come into being, due to constraints on both politics and access. In summary, there would be a stable (if not stagnant) small set of competing networks domestically, with global networks never coming into being.

Many-to-One Information Transfer

This portion of the scenario seems to be unfolding in the immediate present. Remote sensing satellites recently have been plagued by technical failures in the skies and have been the victims of the 1986 launch disasters as well. With a very small public network in place, even minor losses are devastating. As matters stand now, an appreciable gap in the sensing program of the United States appears inevitable.

The private network—military reconnaissance satellites—has also been crippled by the gap in launch capability. Long before SDI is enough of a reality to threaten to absorb all launch capacity, the need for spy satellites is already beginning to usurp commercial payloads.

The proposed commercial takeover of the LANDSAT system has already faltered, perhaps irrevocably, with the United States placing grudging reliance on the French SPOT system. The combination of private-sector failure and national security pressure appear to make militarization of these efforts inevitable.

In addition, developing countries' complaints of exploitation of natural resources are being joined for the first time by industrialized nations with similar fears. In both cases these fears are also tinged with apprehension over private, rather than national, espionage, now possible on a wide scale. If the militarization scenario is a modest setback for broadcasting, and somewhat worse for telephony, it sounds the death-knell for commercial earth observation.

AN ECONOMIC SCENARIO: COMMERCIALIZATION, COMPETITION, AND CONFUSION

One-to-One Information Transfer

If commercial launches slowly become available, the competition to AT&T in domestic telephony will heat up, but still not to the point of dislodging the company from its dominance in the long-distance market. It is very possible that AT&T's expected aggressive reentry into the satellite field would reinforce its advantage, while its would-be competitors are hampered by their late start (and, for the regional Bell companies, by regulatory fetters as well).

Instead, such competitors may turn to a greater reliance on terrestrial fiberoptic networks, and AT&T's own digital transatlantic cable will be its own biggest rival. In general, the telephony world will remain quite centralized, with little opportunity for new private networks or new services, and with greater difficulties for interconnection.

On the international side, potential competition, at least on the transatlantic routes, will be denied on economic grounds, and an allotment plan for geostationary orbit will have a chilling effect on new developments there also. Intelsat and the ITU will still remain viable, however. For one-to-one information transfer, this scenario seems very much like indefinite continuation of business as usual.

One-to-Many Information Transfer

As before, the broadcasting aspects of the scenario are a tepid version of the telephony effects, except that the major cable operators are more likely to gain ground on their national network counterparts. Cost considerations, most associated with the delay in access to orbit, will prompt a greater emphasis on fiberoptic terrestrial networks. But only the major cable operators will be able to compete in the sky or on the ground, while the fringe networks will wither on the vine; the range of the total market will increase, but only modestly.

Primary economic issues will center more on signal ownership problems, with signal piracy, scrambling requirements, and rebroadcasting restrictions all serving as limits to growth. These limits, rather than political ones, will similarly constrain any expansion of international services, although a true global network will move somewhat closer to actuality.

Many-to-One Information Transfer

Remote sensing will remain commercialized rather than militarized, but commercial involvement will be in name only: massive governmental support in terms of both direct and indirect subsidies. A vicious cycle of limits will ensue: with niggardly governmental funding, a commercialization program will move forward only slowly and be unable to offer enhanced services; without growth or novel products, the operation will not be able to generate sufficient volume to become profitable; at the same time, governmental control increases to protect the original investment, with any special services that evolve being directed toward governmental needs; still more private users are driven away; the commercial project requires still greater subsidies; and so on, as the cycle repeats itself until the operation becomes a federal white elephant.

Some modest use of remote sensing data goes on with the two existing systems, Eosat and SPOT, but planned systems are mothballed in the face of the perceived expenses. Nearly all such users will be other governments rather than private actors, and public access to data will be as or more restricted than it is in the present.

A TECHNOLOGICAL SCENARIO: PRIVATE OWNERSHIP, DIGITIZATION, AND DECENTRALIZATION

One-to-One Information Transfer

In this technological scenario, the downward cycle described for earth observation above is reversed for telephony. Greater volume will bring greater use, and satellite communications will finally come into its own. Only two technological assumptions are needed to bring about this case. The first, a quick return to cheap launches, is highly problematical. The second, the dominance of digital satellites, already exists and needs only to be implemented.

On the other hand, the implicit political assumptions behind a decentralization scenario are enormous but, if the technological gains are large enough, they will outstrip national ability or desire to regulate them. In a digitized, low-cost, high-volume world, satellites will not be competitors to fiberoptics, but both forms of transmission will be able to communicate freely to each other in integrated, hybrid networks. Mobile telephony will only be the leading edge of almost completely personalized telecommunications, with AT&T a significant but no longer dominant player.

Internationally, as the size of the pie is perceived to be growing, fears of being denied access will be ameliorated. In particular, availability of geostationary orbit slots will increase several times over due to technical fixes, and will successfully defuse tensions at the ITU. For all three types of systems, this scenario is the realization of a "let a thousand flowers bloom" philosophy.

One-to-Many Information Transfer

Once again, the national networks are locked into the same future as AT&T. The challenge they are now experiencing from superstations and the major cable operators is but the precursor of an almost totally decentralized broadcasting world. The slogan of this world will be "a dish on every garage," with scrambling technology being effective and inexpensive enough to guarantee that every signal has a home. Every fringe operator will be able to maintain a system, with transmission and reception costs being low enough to afford even a minuscule audience.

On the international front, direct-broadcasting satellites will become a reality, either with the help of, or even despite, regulation if broadcasting power is sufficient. Similarly, a wide enough range (both as to variety and power) of transmissions will ensure that the first truly worldwide networks come into existence. For one-to-many information transfer the technological scenario is the advent of the long-heralded global village.

Many-to-One Information Transfer

Private use, yet alone operation, of remote sensing systems is the aspect to this technological scenario that is most removed from the current world, but is just as susceptible to change once the logjam of volume of use is broken. The sheer volume of data that is generated is so overwhelming that the volume of use necessary to maintain it may seem impossible, but this is only because the variety and volume of uses that may occur is also unforeseen. Explosive growth and the low-cost access it enables are also the best antidote to objections of invasions of sovereignty. At the same time, widespread distribution of timely information is a stabilizing force in world affairs.

News agencies will probably be the first private operators (starting as lessors?) of earth observation networks, initiating a trend of specialized systems that need not cover all areas of the earth for all types of radiation. The proposed directional and emergency uses will follow soon after, with educational and entertainment applications not far behind. Additional technical improvements, already in hand, employ new optical and mathematical picture-enhancing techniques to carry applications further still.

When Benjamin Franklin was asked the use of one of his inventions, he replied, "What is the use of a newborn child?" Earth observations networks hold the same limitless promise under this scenario.

CONCLUSION

Some readers may find the political scenario described here too bleak, the economic scenario too bland, and the technological scenario too naive. In the author's opinion, all are plausible, though not equally so, and contain elements of the possible for better or worse. Over time—enough time—the author also believes that some elements of the technological scenario will be successful, dominate countervailing political and economic realities, and change the way the world works. As for the all too real, short-term possibilities for disaster, an optimist is someone who thinks the future is uncertain.

Appendix A
Directory of Publications

AAS DIRECTORY
American Astronautical Society
6060 Tower Court
Alexandria, VA 22304
Biannual: $30
George Cranston, editor 703/751-7721

Directory of the American Astronautical Society.

AAS GODDARD MEMORIAL SYMPOSIA
Univelt, Inc.
P.O. Box 28130
San Diego, CA 92128
Annual
H. Jacobs, editor/publisher 619/716-4005

General coverage of space programs, astronautics, and related fields.

AAS HISTORY SERIES
Univelt, Inc.
P.O. Box 28130
San Diego, CA 92128
Irregular: $14
H. Jacobs, editor/publisher 619/746-4005

History of aeronautics, astronautics, space flight, and related fields. Contains IAA and scientific abstracts.

AAS MICROFICHE SERIES
Univelt, Inc.
P.O. Box 28130
San Diego, CA 92128
Quarterly
H. Jacobs, editor/publisher 619/746-4005

Covers astronautics and related aerospace programs.

AAS/DGLR CONFERENCE PROCEEDINGS
Univelt, Inc.
P.O. Box 28130
San Diego, CA 92128
Biannual
H. Jacobs, editor/publisher 619/746-4005

Deals with space shuttle/space lab ventures and promotes European-United States cooperative space programs.

ACTA ASTRONAUTICA
Pergamon Press, Inc.
Maxwell House, Fairview Park
Elmsford, NY 10523
Monthly: $450
Per issue: $37.50
Luigi Napolitano, editor 914/592-7700

Current issues in propulsion, astrodynamics, guidance, space flight, and atmospheric and space physics.

ADVANCES IN THE ASTRONAUTICAL SCIENCES
Univelt, Inc.
P.O. Box 28130
San Diego, CA 92128
Quarterly
H. Jacobs, editor/publisher 619/746-4005

Deals with topics on space programs and astronautics, technical abstracts, and meetings of the AAS.

AEROSPACE
Aerospace Industries Association of America
1725 DeSales Street, NW
Washington, DC 20036
Quarterly: Free
John Loosbrack, editor 202/429-4600

Covers the relationship between aerospace and technology, trade, defense, and space exploration.

AEROSPACE AMERICA
American Institute of Aeronautics
1633 Broadway
New York, NY 10019
Monthly: $51
Per issue: $10
John Newbawer, editor 212/581-4300

Engineering and management topics in astronautics and aeronautics.

AEROSPACE COMPANIES
Defense Marketing Services, Inc.
100 Northfield Street
Greenwich, Ct 06830
Monthly: $750
Richard Ford, editor 203/661-7800

Sales, research and development efforts, and continuing programs of the top aerospace/defense companies.

AEROSPACE FACTS AND FIGURES
McGraw-Hill
1221 Avenue of the Americas
New York, NY 10020
Annual: $10.95
Janet Martinusen, editor 212/512-2123

Provides a statistical and anayltical overview of the United States aerospace industry.

AEROSPACE HISTORIAN
Air Force Historical Foundation
Kansas State University
Manhattan, KS 66506
Quarterly: $25
Per issue: $10
Robin Higham, editor 913/532-6733

Studies on the traditions and history of aerospace power.

AIA COMMITTEE DIRECTORY
Aerospace Industries Association of America
1725 DeSales Street, NW
Washington, DC 20036
Annual 202/347-2315

Directory of the Aerospace Industries Association of America. Available to members only.

AIAA JOURNAL
American Institute of Aeronautics & Astronautics
1633 Broadway
New York, NY 10019
Monthly: $79
Per issue: $10
George W. Sutton, editor 212/581-4300

Covers new technical and exploratory developments in jet and rocket propulsion, space and atmospheric flight, and other technical engineering areas.

ANNALES DES TELECOMMUNICATIONS
540 F. Centre National D'Etudes des Telecommunications
38-40 Rue du General Leclerc
Moulineaux, France
Bimonthly

Mathematical and scientific abstracts for the telecommunications industries.

ANNIVERSARY CONFERENCES OF THE AAS
Univelt, Inc.
P.O. Box 28130
San Diego, CA 92128
Annual
H. Jacobs, editor 619/746-4005

Covers AAS conferences on space policy, space development, and the space program in general.

ASTROPHYSICS & SPACE SCIENCE
D. Reidel Publishing Company
P.O. Box 17
Dordrecht, The Netherlands 3300 AA
Irregular: $113.50
Zdenek Kopal, editor

A highly technical journal stressing topics of current interest as a result of space research, such as optical and radio equipment, rockets, and satellites.

AVIATION QUARTERLY
Aviation Quarterly Publishers
2512 Hawey Avenue
Dallas, TX 75235
Quarterly: $58
Per issue: $14.95
Brad Bierman, editor 214/423-8516

Well-illustrated, comprehensive source on aviation history.

AVIATION WEEK & SPACE TECHNOLOGY
McGraw-Hill
1221 Avenue of the Americas
New York, NY 10020
Weekly: $39
Per issue: $3
Robert B. Hotz, editor 212/512-4971

A news and features weekly covering the aerospace industry.

BRITISH INTERPLANETARY SOCIETY JOURNAL
British Interplanetary Society
27-29 South Lambeth Road
London, SW8 1SZ, U.K.
Monthly: $115
A. Wilson, editor

Authoritative journal covering space technology, space applications, astronautics history, space and education, and interstellar studies.

CANADIAN AERONAUTICS AND SPACE JOURNAL
Canadian Aeronautics and Space Industry
60-75 Sparks Street
Ottawa, ON K1P5A5, Canada
Quarterly: $30
Per issue: $5
A.A. Buchanan, editor 613/234-0191

Technical papers on aerospace science and engineering subjects.

COMMUNICATIONS TECHNOLOGY IMPACT
Elsevier International Bulletins
52 Vanderbilt Avenue
New York, NY 10017
Monthly: $190
Mick Ennis, editor 212/370-5520

International bulletin for publishing and information handling organizations, containing book reviews, charts, illustrations, and abstracts.

COMSAT TECHNICAL REVIEW
Communications Satellite Corporation
22300 Comsat Drive
Clarksburg, MD 20871
Semiannual: $10
Barbara Wassel, editor 301/428-4000

Scientific, computer, electronics, and physics abstracts that relate to satellite communications.

COSMIC RESEARCH
Plenum Publishing Corp.
233 Spring Street
New York, NY 10013
Bimonthly: $485
L.I. Sedov, editor 212/620-8000

Translation of Kosmicheskie Issledovaniya of the Academy of Sciences, USSR.

EARTH-ORIENTED APPLICATIONS OF SPACE TECHNOLOGY
Pergamon Press, Inc.
Maxwell House, Fairview Park
Elmsford, NY 10523
Quarterly: $120
Per issue: $30
Luigi Napolitano, editor 914/592-7700

Contains articles and studies on the common interests of technology suppliers, developers, and users of space technology.

FLIGHT INTERNATIONAL
Business Press International
205 East 42nd Street
New York, NY 10017
Weekly: $30.55
David Mason, editor 212/867-2080

Deals with issues concerning government, military, and commercial avionics and space flight.

FLIGHT INTERNATIONAL
IPC Transport Press, Ltd.
Dorset House, Stamford Street
London, SE1 9LU, U.K.
Weekly: $78
J.M. Ramsden, editor

Deals with British aerospace events and world news concerning space flight and avionics.

INTERNATIONAL AEROSPACE ABSTRACTS & CUMULATED INDEX
American Institute of Aeronautics & Astronautics
1633 Broadway
New York: NY 10019
Biweekly: $475
Per issue: $25
Irene Bogolubsky, editor 212/581-4300

Index and abstracts of worldwide scientific and technical literature of the aerospace industry.

INTERNATIONAL SPACE BUSINESS REVIEW
Media Dimensions, Inc.
Box 1121 Gracie Station
New York, NY 10028
Irregular (?): $50
Stanley Goldstein, publisher 212/533-7481

Covers topics concerning the commercialization of space.

JOURNAL OF ASTRONAUTICAL SCIENCES
American Astronautical Society, Inc.
6060 Tower Court
Alexandria, VA 22304
Quarterly: $75
Per issue: $18.75
Herbert Rauch, editor 703/751-7721

Deals with topics on the astronautical sciences and contains the International Aerospace abstracts.

JOURNAL OF SPACECRAFT AND ROCKETS
American Institute of Aeronautics & Astronautics
1633 Broadway
New York, NY 10019
Bimonthly: $90
R.H. Woodward Waesche, editor 212/581-4300

Contains original research papers on current advances in space sciences, and missile technologies and applications.

L'INFORMATIQUE BULLETIN
Assoc. Int'l. D'Histoire des Telecommunications
27 Rue Charlot
Paris, France 75003
Per issue: 40f

NEWS BULLETIN
Association for Unmanned Vehicle Systems
1133 15th Street, NW
Washington, DC 20005
Quarterly: 202/429-9440

Covers the development and manufacture of space vehicles, weapons systems, and guidance controls.

PLANETARY & SPACE SCIENCE
Pergamon Press, Inc.
Maxwell House, Fairview Park
Elmsford, NY 10523
Monthly: $400
Per issue: $33.50
D.R. Bates, editor 914/592-7700

This four-color journal presents research papers on the earth and its atmosphere, and interplanetary space, and publishes reviews of books on related subjects.

PROGRESS IN AEROSPACE SCIENCES
Pergamon Press, Inc.
Maxwell House, Fairview Park
Elmsford, NY 10523
Quarterly: $110
Per issue: $27.50
J.P. Finley, editor 914/592-7700

Presents commissioned review articles designed to be of broad interest and use to the research establishment, industry, and universities. Also runs book reviews on related subjects.

SAT-GUIDE, CABLE'S SATELLITE MAGAZINE
Commtek Publishing Company
Box 2700-E
Hailey, ID 83333
Monthly: $36
Per issue: $5
Terrance Stanton, editor 208/788-4936

Contains technical and business articles for the satellite and cable television SMATV, MDS, and DBS.

SATELLITE AGE
Martin Roberts & Associates, Inc.
Box 5254
Beverly Hills, CA 90210
Quarterly: $20
Robert Loper, editor 213/273-0381

Newsletter for the satellite communications industry.

SATELLITE AUDIO REPORT
Waters Information Services
Security Mutual Building, Suite 322
Binghamton, NY 13901
Monthly: $185
Dennis Waters, editor 607/770-1945

Newsletter covers articles concerning the use of satellites to distribute audio signals.

SATELLITE BUSINESS
Video Publishing Company
Box 2772
Palm Springs, CA 92263
Monthly: $36
Steven Tolin, editor 619/323-2000

General business data and news on all phases of the industry.

SATELLITE COMMUNICATIONS
Cardiff Publishing Company
6530 South Yosemite
Englewood, CO 80111
Monthly: $27
Guy Stephens, editor 303/694-1522

SATELLITE DEALER
Commtek Publishing Company
Box 2700-E
Hailey, ID 83333
Monthly: $4
Ron Rudolph, editor 208/788-4936

Devoted to the home satellite industry, covering installation, marketing, programming, and legal issues.

SATELLITE DIRECTORY
Phillips Publishing, Inc.
7315 Wisconsin Avenue
Bethesda, MD 20814
Annual: $95
Dr. Mark Kimmel, editor 301/986-0666

Supplies information on satellite companies, consultants, programmers, and earth stations.

SATELLITE NEWS
Phillips Publishing, Inc.
7315 Wisconsin Avenue
Bethesda, MD 20814
Weekly: $327
Per issue: $7
William Conner, editor 301/986-0666

Newsletter covering marketing, management, and regulation in the satellite industry.

SATELLITE ORBIT
Commtek Publishing Company
Box 2700-E
Hailey, ID 83333
Monthly: $48
Per issue: $5
Bruce Kinnaird, editor 208/788-4936

Lists and updates of all available satellite television services.

SATELLITE TELECOMMUNICATIONS NEWSLETTERS
Martin Roberts & Associates, Inc.
P.O. Box 5254
Beverly Hills, CA 90210
Martin Roberts, publisher

SATELLITE TRANSPONDER SUPPLY/DEMAND ANALYSIS
Telestrategies
6842 Elm Street 102
McLean, VA 22101
Annual (?): $595
703/734-7050

National directory of suppliers.

SATELLITE WEEK
Television Digest, Inc.
1836 Jefferson Place, NW
Washington, DC 20036
Weekly: $327
Jonathan Miller, editor 202/872-9200

Authoritative newsletter covers satellite communications industries and related fields.

SCIENCE AND TECHNOLOGY
Univelt, Inc.
P.O. Box 28130
San Diego, CA 92128
Quarterly
H. Jacobs, editor/publisher 619/746-4005

Presents monographs, proceedings, and abstracts on space science engineering and programs.

SCIENCE AND TECHNOLOGY SERIES
Univelt, Inc.
P.O. Box 28130
San Diego, CA 92128
Quarterly
H. Jacobs, editor/publisher 619/746-4005

Covers astronautics and related areas of space science and engineering, including application to earth problems.

SOVIET AERONAUTICS
Allerton Press, Inc.
150 Fifth Avenue
New York, NY 10011
Quarterly: $245
Per issue: $60
W. Shalof, publisher 212/924-3950

English translation of a Soviet journal of aeronautical science and engineering.

SOVIET AEROSPACE
Space Publications, Inc.
1341 G Street, NW
Washington, DC 20005
Weekly: $200
Norman Baker, editor 202/638-0500

Comprehensive coverage of Soviet space development, defense, and foreign policy actions related to this area.

SPACE (USSR)
U.S. Joint Publication Research Service
1000 North Gleb Road
Arlington, VA 22201
Annual: $60
703/557-4630

SPACE COMMERCE BULLETIN
Television Digest, Inc.
1836 Jefferson Place, NW
Washington, DC 20036
Biweekly: $197
Albert Warren, editor 202/872-9200

Newsletter contains primarily business news for satellite communications.

SPACE COMMUNICATIONS & BROADCASTING
Elsevier Science Publishers, B.V.
Molenwerf
IPO Box 211, 1000AE
Amsterdam, The Netherlands
Bimonthly: $88
B.L. Herden, editor

International English-language journal containing illustrations, indexes, and abstracts.

SPACE LETTER
Callahan Publications
P.O. Box 3751
Washington, DC 20007
Semimonthly: $125
Vincent F. Callahan, Jr., editor/publisher 703/356-1925

Presents contract and procurement data for contractors and subcontractors to NASA, plus book reviews on related subjects.

SPACE MANUFACTURING
Univelt, Inc.
P.O. Box 28130
San Diego, CA 92128
Biannual
H. Jacobs, editor 619/746-4005

Covers the biomedical, social, economic, and international aspects of space manufacturing.

SPACE PRESS
Vernuccio Publications
645 West End Avenue
New York, NY 10025
Monthly: $15
Per issue: $1.25
Frank Vernuccio, Jr., editor 212/724-5919

Provides comprehensive international coverage of news of outer space developments.

SPACE R&D ALERT
Aerospace Communications
350 Cabrini Boulevard
New York, NY 10040
Biweekly: $195
Jeffrey Manber, editor 212/927-8919

Newsletter for the business and financial communities provides coverage of new R&D developments in space industries.

SPACE WORLD
Palmer Publications, Inc.
P.O. Box 296
Amherst, WI 54406
Monthly: $18
Per issue: $2.50
Leonard David, editor 715/824-3214

Deals with National and international space exploration.

TELCOM HIGHLIGHTS
Communications & Marketing Systems
Box 237
Midland, NJ 07432
Weekly: $220
P. Gonsalves, editor

Deals with the entire telecommunications and broadcasting industries.

TELECOMMUNICATIONS POLICY, R&D
U.S. Joint Publication Research Service
1000 North Gleb Road
Arlington, VA 22201
Weekly: $200
Per issue: $5
703/557-4630

TELECOMMUNICATIONS REPORTS
Business Research Publications, Inc.
817 Broadway
New York, NY 10003
Weekly: $205
Fred W. Henck, editor 212/673-4700

Covers legislative, regulatory, tax, and business developments affecting the domestic and international satellite and telecommunications industries.

TELECOMMUNICATIONS TECHNOLOGY (DIANXIN JISHI)
China Publications Center
Chegongchang XILU 21
P.O. Box 339
Beijing, People's Republic of China
Monthly: $.60

Chinese-text journal covers the latest developments in the governmental telecommunications programs of the People's Republic of China.

TELECOMMUNICATIONS WEEK
Business Research Publications
817 Broadway
New York, NY 10003
Weekly: $175
Len Smedresman, editor 212/637-4700

Provides broad coverage of activities affecting the telecommunications industries.

TMSA AEROSPACE MARKET OUTLOOK
Technical Marketing Society of America
3711 Long Beach Boulevard, Suite 609
Long Beach, CA 20005
Bimonthly: $45
Per issue: $5
213/596-0254

Reviews major aerospace programs and analyzes marketing opportunities for a broad spectrum of suppliers to the aerospace industry.

UTIAS REPORT
University of Toronto
4925 Dufferin Street
Downsview, ON M3H5T6, Canada
Irregular: free
416/667-7700

Scientific research papers pertaining to aerospace engineering and related sciences.

Appendix B
Directory of Associations

AEROSPACE EDUCATION FOUNDATION
1501 Lee Highway
Arlington, VA 22209
Russell E. Dougherty, executive director 703/247-5839
Founded: 1956

Wants to broaden the public's understanding of aerospace developments. Maintains the Aerospace Education Center, which provides a forum for experts in these areas, and sends out information regarding new developments in aerospace and aerospace education.

AEROSPACE ELECTRICAL SOCIETY
P.O. Box 24883, Village Station
Los Angeles, CA
Lloyd P. Appelman, president 714/778-1840
Founded: 1941

Technicians, engineers and management personnel engaged in the development and use of electrical/electronic equipment and systems for air and space craft.

AMERICAN ASTRONAUTICAL SOCIETY
6212-B Old Keene Mill Court
Springfield, VA 22152
George E. Cranston, executive secretary 703/866-0020

Promotes and supports scientific research related to the development of astronautical sciences. Members are researchers, scientists, educators, and other professionals in astronautics and other areas.

AMERICAN INSTITUTE OF AERONAUTICS AND ASTRONAUTICS
1633 Broadway
New York, NY 10019
James J. Hartford, executive director 212/581-4300
Founded: 1963

Concerned with facilitating the interchange of technological information through publications and technical meetings, with *overall* progress in the field and increasing the professional competence of its members.

AMERICAN SPACE FOUNDATION
111 Massachusetts Avenue, NW, Suite 200
Washington, DC 20002
Robert R. Weed, executive director 202/289-2293
Founded: 1981

Supports a strong American space program and lobbies to increase NASA's budget to 1 percent of the total federal budget. This association also wants a permanent lunar base by the year 2000, construction of a permanently manned space station by 1992, and to land people on Mars by the year 2000.

ASSOCIATION OF UNITED STATES MEMBERS OF THE
INTERNATIONAL INSTITUTE OF SPACE LAW
Office of the President
University, MS 38677
Professor Stephen Gorove, president 601/232-7361

Promotes the participation of United States members in the International Institute of Space Law, and contributes to the solution of legal problems arising from the use and exploration of outer space.

DIVIDENDS FROM SPACE
1826 Minnesota Avenue
South Milwaukee, WI 53172
Daniel M. Lentz, chairman 414/337-0847
Founded: 1977

Concerned with honoring individuals who have distinguished themselves in both the written and oral presentations of the benefits of space exploration to the public, and to space exploration experts.

HIGH FRONTIER
1010 Vermont Avenue, NW, Suite 1000
Washington, DC 20005
Lt. General Daniel O. Graham (Ret.), president 212/737-4979
Founded: 1982

Economists, scientists, space engineers, and strategists advocating the use of outer space for nonnuclear commercial and military use by the United States and its allies, and the use of equipment currently in development.

INTERNATIONAL ASSOCIATION OF SATELLITE USERS
AND SUPPLIERS
P.O. Box DD
McLean, VA 22101
A. Fred Dassler, president 703/759-2094
Founded: 1980

An international association of satellite suppliers, users, and related organizations. Examines satellite issues, technology trends, and telecommunications planning strategies for commercial, social, national, or international satellite uses. Advises and interprets actions of the FCC, NTIA, and the U.S. Congress.

INTERNATIONAL TELECOMMUNICATIONS SATELLITE
ORGANIZATION (INTELSAT)
3400 International Drive, NW
Washington, DC 20008
R. Collins, director general 202/944-6800
Founded: 1964

Comprised of governments that adhere to two international telecommunications agreements, this association is concerned with the design, construction, maintenance, and operation of the space segment of the global communications satellite system.

L5 SOCIETY
1060 Elm Street
Tuscon, AZ 85719
Phillip K. Chapman, president 602/622-6351
Founded: 1975

This association's goal is to get tens of thousands of people living and working in space as soon as possible.

NATIONAL AVIONICS SOCIETY
500 East 66th Street
Richfield, MN 55423
John Gera, secretary/treasurer 612/866-8800
Founded: 1973

Technicians, engineers, and manufacturers concerned with promoting educational opportunities in various phases of aerospace safety.

NATIONAL ENVIRONMENT SATELLITE DATA
& INFORMATION SERVICE
3300 Whitehaven Street, NW
Washington, DC 20235
Margaret E. Courdin, deputy assistant administrator 202/634-7318
Founded: 1965

This association is a major line component of the National Oceanic and Atmospheric Association (U.S. Dept. of Commerce) and is responsible for the acquisition, processing, dissemination, and recall of worldwide environmental data. Provides information on products for use by governmental agencies and the international scientific, agricultural, and engineering communities.

NATIONAL SPACE CLUB
655 15th Street, NW, Suite 300
Washington, DC 20005
Rory Maher, executive secretary 202/639-4210
Founded: 1957

Concerned with establishing and maintaining United States space leadership and with advancing peaceful and military applications of space flight and related technologies.

NATIONAL SPACE INSTITUTE
West Wing, Suite 203
Washington, DC 20024
Dr. Glen P. Wilson, executive director 202/484-1111
Founded: 1975

Seeks to provide a forum for public participation to help ensure that the space program is responsive to public priorities. Also dedicated to convincing the nation of the critical need for a growing, progressive American space technology.

PROGRESSIVE SPACE FORUMS
1742 Sacramento Street, #9
San Francisco, CA 94109
Jim Heaphy, president 415/673-1079
Founded: 1979

Supports peaceful uses of, and international cooperation in, space; provides news and analyses of United States and international space activities, and is concerned with the relation of women/organized labor and minorities to space activities.

PUBLIC SERVICE SATELLITE CONSORTIUM
600 Maryland Avenue, Suite 220
Washington, DC 20024
Dr. Louis Bransford, president 202/863-0890
Founded: 1975

Represents the telecommunications interests of universities, legal associations, state agencies, trade associations, religious groups, and labor unions on representation to federal agencies. Has launched Campus Conference Network for college tele-conferences.

SATELLITE OPERATORS & USERS TECHNICAL COMMITTEE
GTE Spacenet Corporation
1700 Old Meadow Road
McLean, VA 22101
Leo Torrezao, secretary 703/848-0233

Made up of owners, operators, users, and consultants of satellite systems, this association disseminates and exchanges technical information in an effort to avoid intersatellite interference. Conducts seminars and training programs, and is developing a database of satellite users.

SATELLITE TELEVISION INDUSTRY ASSOCIATION
c/o Chuck Hewitt
300 North Washington Street, Suite 310
Alexandria, VA 22314
Chuck Hewitt, executive vice president 703/549-6990
Founded: 1980

The association hopes to promote public interest in satellite communications, eliminate misconceptions about the use of private earth stations, and establish the rights of private earth stations users. Members are owners/operators of satellite receiving terminals and manufacturers of equipment for satellite earth stations.

SOCIETY OF SATELLITE PROFESSIONALS
P.O. Box 7154
McLean, VA 22106
Elizabeth Harrington, administrator 703/356-3787
Founded: 1983

This association, whose members are college graduates with at least three years' experience in the satellite industry, promotes professional development in the field of satellite applications and hopes to develop a global network of associations. Has already created an international council of advisors in the field of satellite communications.

SPACE STUDIES INSTITUTE
P.O. Box 82
Princeton, NJ 08540
Dr. Gerald K. O'Neill, president 609/921-0377
Founded: 1977

Promotes manufacturing, research, and human exploration of space, and conducts educational programs to further that end.

TECHNICAL INFORMATION SERVICE OF AMERICA
American Institute of Aero & Astronautics
555 West 57th Street
New York, NY 10019
Barbara Lawrence, administrator 212/247-6500

Devoted to the retrieval of scientific publications through its abstracting and indexing services, and extensive technical library and computer services.

U.S. SPACE EDUCATION ASSOCIATION
746 Turnpike Road
Elizabethtown, PA 17022
Stephen M. Cobaugh, international president 717/367-3265
Founded: 1973

Concerned with stimulating public awareness of how mankind benefits from a viable and expanded space program. Maintains a staff of journalists and photographers who cover aerospace events.

UNIVERSITIES SPACE RESEARCH ASSOCIATION
311 American City Building
Columbia, MD 21044
W.D. Cummings, executive director 301/730-2656
Founded: 1969

An international consortium of universities concerned with fostering cooperation among universities, other research organizations, and the U.S. government to advance space research. Manages the Lunar and Planetary Institute in Houston, Texas.

NOTES

CHAPTER 1

1. Arthur C. Clarke, "Extra-Terrestrial Relays," *Wireless World* 51 (October 1984): 305. It has been reprinted frequently, most often in Clarke, *Ascent to Orbit* (1984).

2. Arthur C. Clarke, *Voices from the Sky* (New York: Harper & Row, 1960), 119.

3. Edward W. Plowman, *Space, Earth, and Communication* (Westport, CT: Greenwood Press, 1984), 3.

4. Walter A. McDougall, *The Heavens and the Earth* (New York: Basic Books, 1985), 101.

5. Ibid.

6. Ibid., 102.

7. Stanley Leinwoll, *From Spark to Satellite: A History of Radio Communication* (New York: Scribner, 1977), 190.

8. McDougall, 115.

9. Ibid., 116.

10. "Ownership of the skies" issues are discussed further in chapter 7.

11. McDougall, 118–21; Schichtle, Cass, *The National Space Program: From the Fifties into the Eighties* (Washington, DC: National Defense University, 1983), 44.

12. Delbert Smith, *Communications via Satellite: A Vision in Retrospect* (Boulder, CO: Westview, 1976), 119–28.

13. Leinwoll, 189.

14. McDougall, 119–34.

15. Arthur C. Clarke, *The Promise of Space,* (New York: Harper & Row, 1968), 72.

16. Leinwoll, 190.

17. McDougall, 195–209.

18. Ibid., 219–24.

19. Clarke, *The Promise of Space,* 90–1.

20. Leinwoll, 190–93.

21. McDougall, 329.

22. McDougall, 426–29.

23. McDougall, 354–58.

24. S. Houston Lay, and Howard J. Taubenfeld, *The Law Relating to Activities of Man in Space* (Chicago: University of Chicago Press, 1970), 119–23; see also chapter 5.

25. Leinwoll, 199–202.

26. Lay and Taubenfeld, 123–33; see also chapter 5.

27. Leinwoll, 202.

28. See the deregulation discussion in chapter 5.

29. Abraham Schnapf, *Communications Satellites: Overview and Options for Broadcasters* (New York: American Management Association, 1982), 12–13.

30. B.C. Blevis, "Satellite Communications: A Canadian Perspective," in *Telecommunications in the Year 2000: National and International Perspectives,* edited by Indu B. Singh (Norwood, NJ: Ablex Publishing, 1983), 20–3.

31. J.E.S. Fawcett, *Outer Space: New Challenges to Law and Policy* (Oxford, England: Clarendon Press, 1984), 45–6.

32. Ibid., 81.

33. Alex Roland, "Triumph or Turkey?," *Discover,* November 1985, 29.

34. Kimberly Godwin, "The Proposed Orion and ISI Transatlantic Satellite Systems: A Challenge to the Status Quo," *Jurimetrics* 24 (1984): 297.

35. Jurgen Hausler, and Georg Simonis, "Underdevelopment via Satellite," in *People in Space,* edited by James Katz (New Brunswick, NJ: Transaction Books, 1985), 110–28.

36. Inmarsat is discussed in more detail in chapter 5.

37. 70 FCC 2d 1460, Docket 78-374 (1979).

38. McDougall, 428–29; see also chapter 6.

39. Mark Long, and Jeffrey Keating, *The World of Satellite Television* (Summertown, TN: Book Publishing Co., 1985), 135–36.

40. Edward R. Finch, and Amanda Moore, *Astrobusiness* (New York: Praeger, 1985), 25–30; see also chapter 6.

41. 90 FCC 2d 1159, Docket 80-634 (1982).

42. "FCC Opens the Skies to DBS," *Broadcasting,* 28 June 1982, 27; see also chapter 3.

43. Militarization issues are considered in chapter 9.

44. Long, 29–32, 179–82.

45. Godwin, 308–10. For general deregulation trends, see chapter 5.

46. See the discussion of the insurance industry in chapter 8.

47. The results of the SpaceWARC are treated in chapter 8 as well.

48. "Shuttle Inquiry Is Forging Ahead but Delay Seems Likely," *Space Commerce Bulletin,* 14 February 1986, 2–3.

49. For those requiring such a background, see Emanuel Fthenakis, *Manual of Satellite Communications* (New York: McGraw-Hill, 1984).

50. Herbert Dordick, Helen Bradley, and Burt Nanus, *The Emerging Network Marketplace* (Norwood, NJ: Ablex Publishing, 1981), 26–27.

51. Plowman, 28–29.

52. Ibid., 25–27.

53. Whether there can be ownership of a position in geostationary orbit is discussed in chapter 8.

54. Spacing and registration of orbits and frequencies with the ITU are discussed in chapter 5.

55. Fthenakis, 40–67.

CHAPTER 2

1. Martin Clifford, *Your Telephone: Operation, Selection and Installation* (Indianapolis, IN: H. W. Sams, 1983), 270–74.

2. Eric Lerner, "Designing Communications Satellites: Intelsat VI and Aussat," *Aerospace America* 5 (May 1985): 93.

3. John Bellamy, *Digital Telephony* (New York: Wiley, 1982), 31.

4. Ibid., 64–75.

5. Ithiel de Sola Poole, *Technologies of Freedom* (Cambridge, MA: Harvard University Press, 1983), 28–30.

6. Ibid., 27.

7. Ibid., 30–35.

8. John Brooks, *Telephone: The First Hundred Years* (New York: Harper & Row, 1976), 180–84.

9. See the historical discussion in chapter 3.

10. See the historical discussion in chapter 1.

11. Brooks, 273–74.

12. Ibid., 275.

13. See the discussion of ComSat in chapter 5.

14. Brooks, 277.

15. Gerald Brock, *The Telecommunications Industry: The Dynamics of Market Structure* (Cambridge: Harvard University Press, 1981), 259–60.

16. Brooks, 265–66.

17. See discussion of current Intelsat controversies in chapter 8.

18. Dan Schiller, *Telematics and Government* (Norwood, NJ: Ablex Publishing, 1982), 48–50.

19. Thomas Taylor, "Telecommunications and Public Safety," in *Telecommunications and Productivity*, edited by Mitchell Moss (Reading, MA: Addison-Wesley, 1981), 280–88.

20. Joseph Pelton, and Robert Filep, "Tele-Education by Satellite," in *Toward International Tele-Education*, edited by Wilbur Blume and Paul Schneller (Boulder, CO: Westview, 1984), 149–88.

21. Rudy Bretz, *Media for Interactive Communication* (Beverly Hills, CA: Sage Publications, 1983), 72–79.

22. Ibid., 81–82.

23. Ibid., 87–88.

24. Ibid., 88–92.

25. Dimitris Chorafas, *Telephony Today and Tomorrow* (Englewood Cliffs, NJ: Prentice-Hall), 202–4.

26. Bellamy, 359–92.

27. James Ott, "Instant Delivery," *Commercial Space* 1 (Winter 1986): 66; Calvin Sims, "Federal Express to End Electronic Mail Service," *New York Times,* 30 September 1986, D1.

28. Robert Rovell, and Louis Cuccia, "A New Wave of Communications Satellites," *Aerospace America* 3 (March 1984): 43.

29. Clifford, 274.

30. Bretz, 40–42.

31. Ibid., 42–54.

32. Loy Singleton, *Telecommunications in the Information Age* (Cambridge, MA: Ballinger Publishing Co., 1983), 179–88.

33. Ibid., 115–26.

34. Sally Smith, "Two-Way Cable TV Falters," *New York Times,* 28 March 1984, C25.

35. Edward Slate, and John Popko, "The Next Five Years in Communications," *Telecommunications,* January 1986, 49; and Samuel Simon, *After Divestiture* (White Plains, NY: Knowledge Industry Publications, 1985), 121–25.

36. George Bolling, *AT&T: Aftermath of Antitrust* (Washington, DC: National Defense University, 1983), 111.

37. For technical details, see Walter Bolter, et al., *The Transition to Competition* (1984), 211–342.

38. Bolling, 111–12.

39. Clifford, 294–300.

40. Richard Anglin, "Mobile Satellites: The New Business in Space," *International Space Business Review* 1 (June/July 1985): 6. See also Nathaniel Feldman, "Aerospace Mobile Communications: Building the Mass Market," *Aerospace America* 23 (June 1985): 50.

41. P.M. Boudreau, and R.W. Breithaupt, "Canadian MSat Program Moves Out," *Aerospace America* 6 (June 1985): 62.

42. Ahmed Ghais et al., "Broadening Inmarsat Services," *Aerospace America* 23 (June 1985): 66.

43. Kenneth Owen, "Inmarsat Takes to the Air," *Aerospace America* 24 (July 1986): 14.

44. Douglas McGill, "Teleport on Staten Island Envisioned as City's Link to a Bright Future," *New York Times,* 10 September 1983, 25.

45. Ilene Smith, "Fiber Optics: Can Satellites Compete?," *Satellite Communications,* February 1985, 22.

46. Ibid.

47. Ibid.

48. Brenda Maddox, *Beyond Babel* (New York: Simon & Schuster, 1972), 73–75.

49. Rovell and Cuccia, 49.

50. David Sanger, "Phone Group Plans Pacific Cable Link," *New York Times,* 24 February 1986, D4.

51. Helmut Müller, "Intelsat Goes Commercial," *Interavia* 2 (February 1986): 195.

52. Abraham Schnapf, *Communications Satellites: Overview for Broadcasters* (New York: American Management Association, 1982), 27.

53. Ilene Smith, "Hello, Federal," *Satellite Communications*, July 1985, 27.

54. Ott, 67–68.

55. Smith, 28–29.

56. Ibid.

57. Joanne De Lavan, "HP's High Tech Net," *Satellite Communications*, March 1985, 32.

58. Ibid.

59. John Tyson, "Cutting Costs, Boosting Productivity," *Satellite Communications*, November 1985, 39.

60. Eric Schmitt, "Network Keeps Alaska Legislators in Touch," *New York Times*, 20 August 1985, A21.

61. Robin Toner, "Using Satellites to Reach Voters in the Out There," *New York Times*, 16 December 1985, B10.

62. Eric Berg, "British Telecom and AT&T Venture Expected," *New York Times*, 22 July 1985, D21.

CHAPTER 3

1. Stuart Deluca, *Television's Transformation* (San Diego, CA: A.S. Barnes, 1980), 41.

2. Ibid., 39–40.

3. Anthony Easton, *The Satellite TV Handbook* (Indianapolis, IN: H.W. Sams, 1983), 19.

4. Deluca, 30–2.

5. Ibid., 13–20.

6. Easton, 21.

7. Abraham Schnapf, *Communications Satellites: Overview and Options for Broadcasters* (New York: American Management Association, 1982), 11.

8. Schnapf, 7–8.

9. Deluca, 103–11.

10. Ibid., 107.

11. Ibid., 131–32.

12. Ibid., 180–81.

13. See the discussion of two-way cable services in chapter 2.

14. Local regulatory problems are covered in chapter 8.

15. Easton, 21.

16. Deluca, 203–5.

17. Loy Singleton, *Telecommunications in the Information Age* (Cambridge, MA: Ballinger Publishing Co., 1983), 98–103.

18. Ibid., 49.

19. Ibid., 50–51.

20. David Owen, "Satellite Television," *Atlantic* 255 (6 June 1985): 45.

21. Easton, 81–82.

22. See the discussion of scrambled signals in chapter 7.

23. Mark Long, and Jeffrey Keating, *The World of Satellite Television* (Summertown, TN: Book Publishing Co., 1985), 149.

24. DBS political issues are covered in chapter 8.

25. Deluca, 43.

26. Singleton, 70–71.

27. Easton, 266.

28. Singleton, 76.

29. Reginald Turnill, *Jane's Spaceflight Directory* (London, England: Jane's Publishing Co., 1984), 247–49.

30. Deluca, 217–19.

31. Easton, 209–12.

32. Singleton, 129–30.

33. Deluca, 69–70.

34. Singleton, 132–33.

35. Ibid., 134.

36. Easton, 27; Schnapf, 27.

37. Easton, 41, 165, 191; Schnapf, 22.

38. Deluca, 117; Singleton, 25.

39. Singleton, 52, 71; Easton, 112–18.

40. Deluca, 217.

41. Easton, 47.

42. Ibid.

43. Ibid., 119–20.

44. Long and Keating, 144–45.

45. Easton, 121.

46. Deluca, 218.

47. Easton, 63.

48. Hayward Barnett, "Spreading the Word," *Wire Communications,* September, 1985, 43.

49. Easton, 63; Deluca, 218.

CHAPTER 4

1. Floyd Henderson, and James Merchant, "Microwave Remote Sensing," in *Introduction to Remote Sensing of the Environment,* 2d ed., edited by Benjamin Richason (Dubuque, IO: Kendall-Hunt, 1983), 191, 214.

2. See the discussion of sensed states' rights in chapter 8.

3. Benjamin Richason, "Landsat Platforms, Systems, Images, and Image Interpretation," in *Introduction to the Remote Sensing of the Environment,* 2d ed., edited by Benjamin Richason (Dubuque, IO: Kendall-Hunt, 1983), 130, 135–38; Office of Technology Assessment (OTA), *Remote Sensing and the Private Sector* (Washington, DC: General Printing Office, 1984), 21.

4. Richason, 147–48; OTA, 21.

5. Rights to information also turn on the raw/enhanced distinction; see chapter 8.

6. Marietta Benko, Willem deGraaff, and Gijsbertha Reijnen, *Space Law in the United Nations* (Netherlands: Martinus Nijhoff, 1985), 3.

7. Ibid., 9–10.

8. OTA, 78.

9. Benko et al., 11.

10. OTA, 36.

11. James Burnett, "Commercial Remote Sensing Data Policy," American Bar Association Forum Committee on Air and Space Law, 2 February 1984.

12. See chapter 6 for the detail on Eosat.

13. Benko, et al., 8–9.

14. Pub. L. 98-166, HR 3222 (1983).

15. "Combining Metasat, Landsat Sensors on Polar Orbiter May Avoid Budget Cuts," *Space Commerce Bulletin,* 4 November 1986, 3.

16. Benko et al., 17.

17. F. Chernoff, and B. Russett, editors, *Arms Control and the Arms Race: Selections from the ABM Treaty, Interim Agreement, Salt II Treaty, and Related Documents* (New York: W. H. Freeman, 1985).

18. OTA, 93.

19. OTA, 96–98.

20. Charles Mohr, "Pentagon Fears Delays on Future Spy Satellites," *New York Times,* 24 February 1986, 136.

21. John Anderson, "Remote Sensing Finds Down-to-Earth Applications," *Commercial Space,* Fall 1985, 70, 73.

22. Floyd Sabins, *Remote Sensing: Principles and Interpretation* (San Francisco: W. H. Freeman, 1978), 301–9.

23. National Commission on Libraries and Information Science, *To Preserve the Sense of Earth From Space* (Washington, DC: General Printing Office, 1984), 12.

24. Sabins, 287–90.

25. Donald Rundquist, and Scott Samson, "Application of Remote Sensing in Agricultural Analysis," in *Introduction to Remote Sensing of the Environment,* 2d ed., edited by Benjamin Richason (Dubuque, IO: Kendall-Hunt, 1983), 317, 328–30.

26. National Commission on Libraries and Information Science, 11–12.

27. William Hilborn, "Application of Remote Sensing in Forestry," in *Introduction to Remote Sensing of the Environment,* 2d ed., edited by Benjamin Richason (Dubuque, IO: Kendall-Hunt, 1983), 338, 348–50.

28. Sandra Blakeslee, "Satellites Assist Pacific Fishermen," *New York Times,* 18 December 1983, I40.

29. John Bassler, and William Carter, "Geodesy Goes High Tech," *Aerospace America* 24 (April 1986): 72.

30. Gary Higgs, and Marvel Lang, "Application of Remote Sensing in Urban Analysis," 378–400; *Introduction to Remote Sensing of the Environment,* 2d ed., edited by Benjamin Richason (Dubuque, IO: Kendall-Hunt, 1983).

31. John Noble Wilford, "Weather Satellite Marks a Birthday," *New York Times,* 2 April 1985, C2.

32. James Gleick, "The Future of Long-Term Broadcasting Is Bright," *New York Times,* 20 October 1985, IV7.

33. Jack Villmow, "Application of Remote Sensing in Weather Analysis," in *Introduction to Remote Sensing of the Environment,* 2d ed., edited by Benjamin Richason (Dubuque, IO: Kendall-Hunt, 1983), 476–97.

34. S.I. Rasoon, "Predicting Earth's Dynamic Changes," *Aerospace America* 24 (January 1986): 78–82.

35. Cass Schichtle, *The National Space Program: From the Fifties into the Eighties* (Washington, DC: National Defense University, 1983), 34.

36. Jean-Louis Magdelenat, "The Controversy over Remote Sensing," in *People in Space,* edited by James Katz (New Brunswick, NJ: Transaction Books, 1985), 129, 130–31.

37. Sabins, 344–48.

38. Kenneth Carstens, Thomas Kind, and Neil Weber, "Application of Remote Sensing in Cultural Resource Management: Archaeology," in *Introduction to Remote Sensing of the Environment,* 2d ed., edited by Benjamin Richason (Dubuque, IO: Kendall-Hunt, 1983), 299, 302.

39. John Noble Wilford, "Spacecraft Detects Sahara's Past," *New York Times,* 26 November 1982, A1.

40. Carstens, et al., 300.

41. Howard Latin, Gary Tannehill, and Robert White, "Remote Sensing Evidence and Environmental Law," *California Law Review* 64 (1976): 1300.

42. Ibid., 1352–53.

43. Ibid., 1351.

44. These details are discussed in chapter 8.

45. Walter Sullivan, "Distress Satellite Loses Orbit," *New York Times,* 14 July 1984, 46.

46. "U.S. Signs a Satellite Pact With 3 Countries," *New York Times,* 18 October 1984, A18.

47. William F. Buckley, "Precision Sailing," *New York Times,* 19 May 1985, VI134, 138.

48. Francis Kane, and John Scheerer, "The Global Role of Navstar," *Aerospace America* 22 (July 1984): 42.

49. Edward Finch, and Amanda Moore, *Astrobusiness* (New York: Praeger, 1985), 4–5.

50. Anderson, 73.

51. Sabins, 364–73.

52. Irvin Molotsky, "Chernobyl and the 'Global Village,'" *New York Times,* 8 May 1986, A22.

53. Thomas Karas, *The New High Ground* (New York: Simon & Schuster, 1983), 114–15.

54. Ibid.

55. Paul Stares, *The Militarization of Space: U.S. Policy, 1945–84* (Ithaca, NY: Cornell University Press, 1985), 14–15.

56. James Bramford, "America's Supersecret Eyes in Space," *New York Times,* 13 January 1985, VI38, 39.

57. James Canan, *War in Space* (New York: Harper & Row, 1982), 102–4.

58. James Fawcett, *Outer Space: New Challenges to Law and Policy* (Oxford, England: Clarendon Press, 1984), 80–1.

59. "Ariane V16 Launch Provides Bright Spot in Recent Space Turmoil," *Space Commerce Bulletin,* 28 February 1986, 7–8.

60. Brendan Greeley, "Commercial Marketing of Landsat Data Begins," *Commercial Space,* Fall 1985, 66, 68.

61. OTA, 104–5, 108–10.

62. OTA, 78.

63. OTA, 138.

64. Anderson, 73.

65. OTA, 118–19.

CHAPTER 5

1. A detailed discussion of alternative launch services is in chapter 6.

2. The project was a West German remote sensing satellite, SPARX, in the summer of 1984.

3. Edward Finch, and Amanda Moore, *Astrobusiness* (New York: Praeger 1985), 24–30, 131.

4. Richard Givens, "Major Policy Options," in *Legal Strategies for Industrial Innovation,* edited by Richard Givens (Colorado Springs, CO: Shepards-McGraw, 1982), 523–40.

5. Eigi, IBM, ComSat, and Aetna, in the original Satellite Business Systems venture; RCA and Hughes in Eosat; Mitsubishi and Ford in Japan Space Communications; and the sharing of facilities by AT&T and the major broadcasters.

6. Internal Revenue Code, Sec. 48 (a) (2), (7).

7. These topics are discussed in chapter 8.

8. Transborder data flow problems are briefly reviewed in chapter 8.

9. The ITU regulations have effectively made this possible by requiring applications for direct broadcast satellites to be made by the receiving state.

10. See, e.g., Waldrog, Mitchell, "Imaging the Earth (II): The Politics of Landsat," *Science* 216 (2 April 1982).

11. See note 5 above.

12. See the technical discussion in chapter 1.

13. Article IX, Treaty on Principles Governing the Activities of States in the Exploration and Use of Outer Space, Including the Moon and Other Celestial Bodies, 18 UST 2410, TIAS 6347, 610 UNTS 205 (1967).

14. See the technical discussion in chapter 1.

15. Finch and Moore, 23–35.

16. James Myers, "Federal Government Regulation of Commercial Operations Using Expendable Launch Vehicles," *Journal of Space Law* 12 (1984): 40–41.

17. Ibid., 48.

18. Presidential Directive/National Security Council NSC-42, 4 July 1982.

19. Exec. Order No. 12465, 49 Fed. Res. 7211 (1984).

20. Commercial Space Launch Act, PL 98-575, 49 USC 2601 (1984).

21. "Commercial Space Transportation: Licensing Process for Commercial Space Launch Activities," Office of the Secretary of Transportation, United States Department of Transportation, OST Docket 42885, Notice 85-3 (2/15/85), 10–11.

22. Ibid., 11–14.

23. Myers, 45–46.

24. Office of the Secretary of Transportation, 14.

25. Commercial Space Launch Act, Sec. 8(a).

26. L. J. Evans, "National Commercial Space Initiatives: Legal Implications," American Bar Association Forum Committee on Air and Space Law, 1 November 1984.

27. PL 73-416, US Statutes at Large (1934).

28. Loy Singleton, *Telecommunications in the Information Age* (Cambridge, MA: Ballinger Publishing Co., 1983), 85.

29. Ithiel de Sola Poole, *Technologies of Freedom* (Cambridge: Harvard University Press, 1983), 49–51.

30. Herbert Marks, "Regulation and Deregulation in the United States and other Countries," *Jurimetrics* 25 (1984): 7–14.

31. Reginald Stuart, "U.S. Limits Local Rules for Satellite Receivers," *New York Times*, 15 January 1986, A16.

32. National Aeronautics and Space Act of 1958, PL 85-568, 72 Stat. 426 (1958).

33. Communications Satellite Act of 1962, PL 87-624, 76 Stat. 419 (1962).

34. Edward W. Plowman, *Space, Earth, and Communication* (Westport, CT: Greenwood Press, 1984), 75.

35. Marks, 14–16.

36. Ibid., 17.

37. Youichi Ito, "Recent Trends in Telecommunications Regulation and Markets in Japan," *Jurimetrics* 25 (1984): 70.

38. Marks, 18.

39. Plowman, 146–49.

40. 18 UST 2410, TIAS 6347.

41. 19 UST 7570, TIAS 6599.

42. 24 UST 2389, TIAS 7762.

43. 28 UST 695, TIAS 8480.

44. Finch and Moore, 75–76.

45. Richard Colino, "Intelsat: Facing the Challenge of Tomorrow," *Columbia Journal of International Affairs* 39 (1985): 129.

46. *Intelsat 1985 Yearly Report* 1985, 11.

47. Joseph Pelton, "The Proliferation of Communications Satellites: Gold Rush in the Clarke Orbit," in *People in Space*, edited by James Katz (New Brunswick, NJ: Transaction Books, 1985), 98, 105–6.

48. Marcia Smith, "U.S. International Space Activities," in *People in Space*, edited by James Katz (New Brunswick, NJ: Transaction Books, 1985), 82, 87.

49. Michel Bourely, "The Contributions Made by International Organizations to the Formation of Space Law," *Journal of Space Law* 10 (1982): 139.

CHAPTER 6

1. Reginald Turnhill, *Jane's Spaceflight Directory* (London: Jane's Publishing Co., 1984), 43 (hereafter *Jane's*).

2. *Space Commerce Bulletin*, 7 November 1986, 7.

3. *The 1986 Satellite Directory* (Baltimore: Phillips Publishing, Inc.), 86–7 (hereafter *Phillips*).

4. *Space Commerce Bulletin*, 19 December 1986, 3.

5. *Jane's*, 27.

6. *Jane's*, 32–33.

7. *Space Commerce Bulletin*, 24 October 1986, 9.

8. *Space Commerce Bulletin*, 9 May 1986, 9.

9. Craig Covault, "Economic Competition," *Commercial Space*, Fall 1985, 18–21.

10. Philip Boffey, "7-Year Schedule is Set in Space Shuttle Tasks," *New York Times*, 4 October 1986, 5.

11. John Wilford, "Test of European Rocket to Delay Launching," *New York Times*, 3 December 1986, A18.

12. William Broad, "Newest Titan Groomed as Rival for Shuttle," *New York Times*, 20 May 1986, C1.

13. "Aerospace Spotlight," *Aerospace America* 24 (October 1986): 1.

14. S. Budiansky, "Business in Space Gets a Boost," *US News & World Report*, 1 September 1986, 60–61.

15. *Space Commerce Bulletin*, 24 October 1986, 3–4.

16. Edward Finch, and Amanda Moore, *Astrobusiness* (New York: Praeger), 32–33.

17. *Space Commerce Bulletin*, 26 September 1986, 3.

18. Keith Mordoff, "Expendable Launchers," *Commercial Space*, Fall 1985, 29–30.

19. *Jane's*, 286.

20. *Space Calendar*, 29 September–5 October 1986, 4.

21. *Space Calendar*, 24 November–30 November 1986, 4.

22. *Space Commerce Bulletin*, 4 July 1986, 5–6.

23. *Space Commerce Bulletin*, 10 October 1986, 2.

24. Joseph Burns, "China's Proud Space Program," *New York Times*, 19 May 1986, D10; *Space Calendar*, 2 June–8 June 1986, 3–4.

25. David Sanger, "Japan Missing a Chance to Be a Leader in Space," *New York Times*, 8 December 1986, D1.

26. John Wilford, "Plans For Space Shuttles Flourish Outside the U.S.," *New York Times*, 12 August 1986, C1.

27. *Space Commerce Bulletin*, 19 December 1986, 3–4.

28. Sanger, David, "Comsat, Contel Plan to Merge," *New York Times*, 30 September 1986, D1.

29. Barnaby Feder, "MCI Struggles to Ward off Deregulation's Sting," *New York Times*, 13 August 1986, III6.

30. Richard Colino, "A Chronicle of Policy and Procedure: The Formulation of the Reagan Administration Policy on International Satellite Telecommunications," *Journal of Space Law* 13 (1985): 105.

31. See the discussion of mobile satellite systems in chapter 2.

32. Calvin Sims, "Federal Express to End Electronic Mail Service," *New York Times*, 30 September 1986, D1.

33. See the discussion of this issue in chapter 8.

34. Timothy Logue, "US Decisions on Pacific Telecommunications Facilities: Letting a Million Circuits Bloom?" *Jurimetrics* 27 (1985): 72–73.

35. Ibid., 73.

36. *Space Commerce Bulletin*, 19 December 1986, 5–6.

37. *Satellite Week*, 10 November 1986, 1.

38. Anthony Easton, *The Satellite TV Handbook* (Indianapolis, IN: H. W. Sams, 1983), 240–245; Mark Long, and Jeffrey Keating, *The World of Satellite Television* (Summertown, TN: Book Publishing Co., 1985), 183–84.

39. Long and Keating, 181–83.

40. See the DBS discussion in chapter 8.

41. LANDSAT's history and the formation of Eosat are discussed in chapter 4.

42. *Space Commerce Bulletin*, 2 January 1987, 6.

43. *Space Commerce Bulletin*, 26 September 1986, 9, and 7 November 1986, 5–6.

44. *Phillips*, 330–31.

45. Pierre Condoin, "Brazil Aims for Self-Sufficiency in Space," *Interavia*, January 1986, 99–101.

46. Bob Jaques, "Canada's Multi-Role Radar Satellite," *Interavia*, April 1986, 433–34.

47. Steven Weisman, "India Carries its Dreams Into Space," *New York Times*, 23 December 1986, C1.

48. *Space Calendar*, 14 October–20 October 1985, 3.

49. *Phillips*, 88.

CHAPTER 7

1. For an NWIO summary, see International Commission for the Study of Communications Problems, *Many Voices, One World* (New York: Unipub, 1980) (The McBride Report).

2. John Wilson, "Computers: When Will the Slump End?" *Business Week*, 21 April 1986, 58.

3. John Bellamy, *Digital Telephony* (New York: Wiley, 1982), 192–94.

4. Anthony Easton, *The Satellite TV Book* (Indianapolis, IN: H. W. Sams, 1983), chapter 8.

5. Joseph Sullivan, "Cordless Phones Raise an Eavesdropping Issue," *New York Times*, 11 March 1986, B3.

6. Ibid.

7. *U.S.* v. *Cotroni*, 527 F. 2D 708 (2d Cir. 1975); *Stowe* v. *DeVoy*, 588 F. 2d 336 (2d Cir. 1978).

8. Samuel Simon, *After Divestiture* (White Plains, NY: Knowledge Industry Publications, 1985), 51–54.

9. Michelle Gouin, and Thomas Cross, *Intelligent Buildings: Strategies for Technology and Architecture* (Homewood, IL: Dow Jones-Irwin, 1985).

10. Jonathan Miller, "Gold Rush in the Sky," *Channels 1986 Field Guide* (1986): 53.

11. David Andrews, "The Legal Challenge Posed by the New Technologies," 24 *Jurimetrics* (1983): 53–54.

12. Reginald Stuart, "TV Groups Reach Pact on Must-Carry Rule," *New York Times,* 28 February 1986, C23.

13. Reginald Stuart, "U.S. Limits Local Rules for Satellite Receivers," *New York Times,* 15 January 1986, A16.

14. Ivor Peterson, "Scrambling of Signals Today Thwarts TV Dish Antennas," *New York Times,* 15 January 1986, A1.

15. Ibid.

16. Ibid.

17. Stuart Deluca, *Television's Transformation* (San Diego, CA: A. S. Barnes, 1980), 189.

18. Ibid., 199.

19. Easton, 203–4.

20. Brendan Greeley, "Commercial Marketing of Landsat Data Begins," *Commercial Space,* Fall 1985, 66.

21. Ibid.

22. William Broad, "Civilians Use Satellite Photos for Spying on Soviet Military," *New York Times,* 7 April 1986, A1.

CHAPTER 8

1. See the discussion of deregulation in chapter 5.

2. "Reaction to White House Space Policy Mixed," *Space Commerce Bulletin,* 29 August 1986, 2; Pat Jefferson, "Congress Delves into ELV's," *Aerospace America* 24 (September 1986): 13.

3. Gerald Boyd, "Aids Say Reagan Will Order NASA to Build Shuttle," *New York Times,* 15 August 1986, A1.

4. "Senate Authorizes Landsat Funding," *Space Commerce Bulletin,* 15 August 1986, 7.

5. David Sanger, "Commercial Launching Shifted to Private Sector," *New York Times,* 16 August 1986, 1.

6. "NASA Would Buy Commercial Launch Services," *Space Commerce Bulletin,* 15 August 1986, 3.

7. James Myers, "Terminating Business Use of the Shuttle," *International Space Business Review,* July/August 1986, 29.

8. Charles Mohr, "Pentagon Fears Delays on Future Spy Satellites," *New York Times,* 24 February 1986, B6.

9. William Broad, "Military Missions to Dominate Role of Space Shuttle," *New York Times*, 24 March 1986, A1.

10. John Wilford, "Plans for Space Shuttles Flourish Outside the US," *New York Times*, 12 August 1986, C1.

11. Robert Brodsky, et al., "Foreign Launch Competition Growing," *Aerospace America* 24 (July 1986): 36.

12. "Satellites Outpace Customers," *New York Times*, 10 April 1984, D1.

13. Barnaby Feder, "Space Shuttle and Commerce," *New York Times*, 11 March 1986, D2, quoting Anthony Cipriano of the Space Commerce Round Table.

14. Wilbur Pritchard, and Robert Nelson, "Challenger Electrifies Transponder Market," *Aerospace America* 24 (May 1986): 9.

15. This and following estimates of insurance losses are taken from Herbert Coleman, "Space Insurance," *Commercial Space*, Fall 1985, 61.

16. "Indonesia May Buy Back Rescued Satellite," *Space Commerce Bulletin*, 18 July 1986, 6.

17. Jeffrey Manber, "Insurers Shy Away From Satellites," *New York Times*, 19 January 1986, VI17.

18. Eric Lerner, "Space Insurance—Who Takes the Risk?" *Aerospace America* 24 (March 1986): 12.

19. "Shuttle Will Refly for Half Price," *Space Business News*, 21 (October 1985): 1.

20. Jerome Simonoff, "The Financial Implications of the Insurance Situation," *International Space Business Review* 1 (June/July 1985): 62.

21. Lerner, 13.

22. Ibid.

23. Simonoff, 63.

24. "Intec Sets up Committee to Consider Private Launch Insurance Solutions," *Space Commerce Bulletin* 15 August 1986, 6.

25. Joel Greenberg, and Carole Gaelick, "Simulation to the Rescue of Space Insurance," *Aerospace America* 24 (August 1986): 62.

26. Craig Dougherty, "A Financial Prospective of Satellite Projects," *International Space Business Review* 1 (June/July 1985): 64.

27. Clayburgh, William, "On Buying Spaceships, Profits, and Dreams," *International Space Business Review* 1 (July/August 1986): 55.

28. The list of United States disasters is taken from an article describing the most recent, "Air Force Blows Up a Test Missile as Flight Over Pacific Goes Awry," *New York Times*, 29 August 1986, A6.

29. William Broad, "Titan Loss May Force Early Use of Shuttle," *New York Times*, 22 April 1986, C1.

30. Judith Miller, "Latest Failure Cripples West's Satellite Ability," *New York Times*, 1 June 1986, A1.

31. James Gleick, "Errant Rocket Destroyed by Ground Control," *New York Times*, 28 August 1986, B11.

32. "Weather Satellite Launched Atop Rebuilt Rocket," *New York Times*, 18 September 1986, A17.

33. David Sanger, "NASA Will Launch Earth Satellite Using its Rockets," *New York Times*, 12 March 1986, A1.

34. "House, Air Force Pursue Separate Paths to Encourage Commercial ELV Market," *Space Commerce Bulletin*, 15 August 1986, 3.

35. Jahn Olushman, "Air Force to Build New Rocket Fleet to Loft Satellites," *New York Times*, 1 August 1986, A1.

36. William Broad, "Newest Titan Groomed as Rival for Shuttle," *New York Times*, 20 May 1986, C1.

37. William Clayburgh, "On Buying Spaceships, Profits, and Dreams," *International Space Business Review* 1 (July/August 1986): 55.

38. Intelsat's history and organization are discussed in chapter 5.

39. The Intelsat challenges continue to be discussed exhaustively. Material here is based on Kimberly Godwin, "The Proposed Orion and ISI Transatlantic Satellite Systems: A Challenge to the Status Quo," *Jurimetrics* 24 (1984): 297; David Leive, "Intelsat In a Changing Telecommunications Environment," *Jurimetrics* 25 (1984): 82; Richard Colino, "A Chronicle of Policy and Procedures: The Formulation of the Reagan Administration Policy on International Satellite Telecommunications," 13 *Journal of Space Law* 13 (1985): 103; Julianne McKenna, "Bypassing Intelsat: Fair Competition or Violation of the Intelsat Agreement," 8 *Fordham Journal of International Law* 8 (1985): 479; Lawrence Caplan, "The Case For and Against Private International Communications Satellite Systems," *Jurimetrics* 26 (1986): 1980; Cheryl Sarreals, "International Telecommunications Satellite Services: The Spirit of Cooperation vs. the Battle for Competition," *Jurimetrics* 26 (1986): 267; and Timothy Logue, "U.S. Decisions on Pacific Telecommunications Facilities: Letting a Million Circuits Bloom?" *Jurimetrics* 27 (1986): 65.

40. Godwin, 304–5.

41. Ibid., 306–7.

42. Colino, 105–6.

43. Logue, 71.

44. Communications Satellite Act of 1962, 47 USC 721 (a)(b) (1979).

45. McKenna, 496–501.

46. Caplan, 194–96.

47. Reginald Stuart, "Intelsat Proves a Tough Foe," *New York Times*, 14 July 1986, D2; *Space Commerce Bulletin*, 19 December 1986, 5–6.

48. Sarreals, 301–4.

49. Ibid., 297–301.

50. Colino, 136–40, 146–53.

51. Michael Gawdun, "Virtual Private Networks," *Telecommunications*, April 1986, 59–61.

52. Walter Bolter, et al., *Telecommunication Policy for the 1980s: The Transition to Competition* (Englewood Cliffs, NJ: Prentice-Hall, 1984), 220.

53. A.M. Rutowski, "The Integrated Services Digital Network: Issues and Options for the Future," *Jurimetrics* 24 (Fall 1983): 19; see also the discussion of digitization in chapter 2.

54. Gerald Brock, *The Telecommunications Industry* (Cambridge: Harvard University Press, 1981), 267–68, 276–77; see also the discussion of the Satellite Business Systems venture at the close of chapter 2.

55. Michael Hashemi, "The Replacement of the World's Largest Private Network," *Telecommunications*, April 1986, 49–52; George Bolling, *AT&T: After-*

math of Antitrust (Washington, DC: National Defense University, 1983): 131–32.

56. John Wilke, "The Small Fry in Phones Are Being Cut Off," *Business Week,* 4 November 1985, 98.

57. Brock, 266–86.

58. Ithiel de Sola Poole, *Technologies of Freedom* (Cambridge: Harvard University Press, 1983), 196–99.

59. Linda Greenhouse, "The Wiretapping Law Needs Some Renovation," *New York Times,* 1 June 1986, IV4.

60. David Burnham, "Loophole in Law Raises Concern About Privacy in Computer Age," *New York Times,* 19 December 1983, A1.

61. R. Lingl, "Risk Allocation in International Interbank Electronic Fund Transfers: CHIPS & SWIFT," *Harvard International Law Journal* 22 (1981): 621, 659.

62. Frank Brauer, "Secure UHF Satellite Links for the '90s," *Aerospace America* 23 (June 1985): 54–57.

63. David Andrews, "The Legal Challenge Posed by the New Technologies," *Jurimetrics* 24 (Fall 1983): 43, 52–53.

64. See the discussion of the Outer Space Treaty in chapter 5.

65. The Declaration was signed by Brazil, Colombia, the Congo, Ecuador, Indonesia, Kenya, Uganda, and Zaire; see Carl Christol, *The Modern International Law of Outer Space* (Elmsford, NY: Pergamon, 1982), 891–96.

66. Ram Jakhu, "The Legal Status of the Geostationary Orbit," *Annals of Air and Space Law* 7 (1982): 349–50.

67. Gregory Staple, "The New World Satellite Order: A Report from Geneva," *American Journal of International Law* 80 (1986): 699.

68. Barbara Waite, and Ford Rowan, "International Communications Law, Part II: Satellite Regulation and the Space WARC," *The International Lawyer* 20 (Winter 1986): 364–65.

69. Walter Sullivan, "New Satellite Technology Will Beam TV to Remote Regions," *New York Times* 30 August 1983, C3.

70. Anthony Easton, *The Satellite TV Handbook* (Indianapolis, IN: H. W. Sams, 1983), 281–82.

71. Fawcett, J.E.S., *Outer Space: New Challenges to Law and Policy* (Oxford, England: Clarendon Press, 1984), 65–77.

72. For a discussion of the recent decision, see Ernest Sanchez, and Diane Mooney, "Current Legal Developments in Direct Broadcasting Satellites," 25 *Jurimetrics* 25 (Winter 1985): 218–19; for an amusing analysis of delaying tactics employed, see John Chapman, and Gabriel Warren, "Direct Broadcast Satellites: The ITU, UN, and the Real World," *Annals of Air and Space Law* 4 (1979): 413.

73. Eric Pace, "TV Curb Backed by UN Assembly," *New York Times,* 11 December 1982, 6.

74. Sanchez and Mooney, 220–21.

75. Andrew Pollack, "Plan for TV by Satellite Falls Apart Over Risks," *New York Times,* 1 December 1984, 31.

76. Peter Boyer, "HBO Piracy Incident Stuns Other Satellite Users," *New York Times,* 29 April 1986.

77. Donald Goldberg, "Captain Midnight, HBO, and World War III," *Mother Jones*, October 1986, 26–29, 48–53.

78. Scrambling issues are discussed in chapter 7.

79. Richard Campagna, "Video and Satellite Transmission Piracy in Latin America: A Survey of Problems, Legal Strategies, and Remedies," *The International Lawyer* 20 (Summer 1986): 961.

80. Fawcett, 30–31.

81. These issues are treated in chapter 5.

82. Jean-Louis Magdelenat, "The Controversy Over Remote Sensing," in *People in Space*, edited by James Katz (New Brunswick, NJ: Transaction Books, 1985), 129–39.

83. Ibid., 135–36.

84. Paul Szasz, "Report on the 25th Session of the Legal Sub-Committee of the UN Committee on the Peaceful Uses of Outer Space, 24 March–11 April 1986," *Journal of Space Law* 14 (1986): 48.

85. "Eosat May Scale Back or Halt Satellite Development Plans," *Space Commerce Bulletin*, 21 November 1986, 4.

86. William Broad, "Satellite Photos Appear to Show Construction of Soviet Space Shuttle Base," *New York Times*, 25 August 1986, A20.

87. William Broad, "Civilians Use Satellite Photos for Spying on Soviet Military," *New York Times*, 7 April 1986, A1.

88. "Remote Sensing Satellites Can Be Useful News Tool, RTNDA Is Told," *Space Commerce Bulletin*, 20 June 1986, 6–7.

89. Peter Glaser, and Mark Brender, "The First Amendment in Space: News Gathering From Satellites," *Issues in Science and Technology*, Fall 1986, 60–67.

BIBLIOGRAPHY

BOOKS

Bellamy, John. *Digital Telephony.* New York: Wiley, 1982.

Benko, Marietta; deGraaff, Willem; and Reijnen, Gijsbertha. *Space Law in the United Nations.* The Netherlands: Martinus Nijhoff, 1985.

Blevis, B. C. "Satellite Communications: A Canadian Perspective." In *Telecommunications in the Year 2000: National and International Perspectives,* edited by Indu B. Singh. Norwood, NJ: Ablex Publishing, 1983.

Bolling, George. *AT&T: Aftermath of Antitrust.* Washington, DC: National Defense University, 1983.

Bolter, Walter, et al., *The Transition to Competition.* Englewood Cliffs, NJ: Prentice-Hall, 1984.

Bretz, Rudy. *Media for Interactive Communication.* Beverly Hills: Sage Publications, 1983.

Brock, Gerald. *The Telecommunications Industry: The Dynamics of Market Structure.* Cambridge: Harvard University Press, 1981.

Brooks, John. *Telephone: The First Hundred Years.* New York: Harper & Row, 1976.

Canan, James. *War in Space.* New York: Harper & Row, 1982.

Carstens, Kenneth; Kind, Thomas; and Weber, Neil. "Application of Remote Sensing in Cultural Resource Management: Archaeology." In *Introduction to Remote Sensing of the Environment,* 2d ed., edited by Benjamin Richason. Dubuque, IO: Kendall-Hunt, 1983.

Chernoff, F. and Russett, B., eds. *Arms Control and the Arms Race: Selections from the ABM Treaty, Interim Agreement, Salt II Treaty, and Related Documents.* New York: W. H. Freeman, 1985.

Chorofas, Dimitris. *Telephony Today and Tomorrow.* Englewood Cliffs, NJ: Prentice-Hall, 1984.

Christol, Carl. *The Modern International Law of Outer Space.* Elmsford, NY: Pergamon, 1982.

Clarke, Arthur C. *Ascent to Orbit.* New York: Wiley, 1984.

Clarke, Arthur, C. *The Promise of Space*. New York: Harper & Row, 1986.

Clarke, Arthur C. *Voices from the Sky*. New York: Harper & Row, 1965.

Clifford, Martin. *Your Telephone: Operation, Selection and Installation*. Indianapolis, IN: H. W. Sams, 1983.

Deluca, Stuart. *Television's Transformation*. San Diego, CA: A. S. Barnes, 1980.

Dordick, Herbert; Bradley, Helen; and Nanus, Burt. *The Emerging Network Marketplace*. Norwood, NJ: Ablex Publishing, 1981.

Easton, Anthony. *The Satellite TV Handbook*. Indianapolis, IN: H. W. Sams, 1983. Pp. 203–4.

Fawcett, James E. *Outer Space: New Challenges to Law and Policy*. Oxford: England, Clarendon Press, 1984.

Finch, Edward, and Moore, Amanda. *Astrobusiness*. New York: Praeger, 1985.

Fthenakis, Emanuel. *Manual of Satellite Communications*. New York: McGraw-Hill, 1984.

Givens, Richard. "Major Policy Options." In *Legal Strategies for Industrial Innovation*, edited by Richard Givens. Colorado Springs, CO: Shepards-McGraw, 1982.

Gouin, Michelle, and Cross, Thomas. *Intelligent Buildings: Strategies for Technology and Architecture*. Homewood, IL: Dow Jones-Irwin, 1985.

Hausler, Jurgen, and Simonis, Georg. "Underdevelopment via Satellite." In *People in Space*, edited by James Katz. New Brunswick, NJ: Transaction Books, 1985. Pp. 110–28.

Henderson, Floyd, and Merchant, James. "Microwave Remote Sensing." In *Introduction to Remote Sensing of the Environment*, 2d ed., edited by Benjamin Richason. Dubuque, IO: Kendall-Hunt, 1983. Pp. 191, 214.

Higgs, Gary, and Lang, Marvel. "Application of Remote Sensing in Urban Analysis." In *Introduction to Remote Sensing of the Environment*, 2d ed., edited by Benjamin Richason. Dubuque, IO: Kendall-Hunt, 1983. Pp. 378–400.

Hilborn, William. "Application of Remote Sensing in Forestry." In *Introduction to Remote Sensing of the Environment*, 2d ed., edited by Benjamin Richason. Dubuque, IO: Kendall-Hunt, 1983. Pp. 338, 348–50.

International Commission for the Study of Communications Problems. *Many Voices, One World*. New York: Unipub, 1980.

Karas, Thomas. *The New High Ground*. New York: Simon & Schuster, 1983.

Lay, S. Houston, and Taubenfeld, Howard J. *The Law Relating to Activities of Man in Space*. Chicago: University of Chicago Press, 1970.

Leinwoll, Stanley. *From Spark to Satellite: A History of Radio Communication*. New York: Scribner, 1977.

Long, Mark, and Keating, Jeffrey. *The World of Satellite Television*. Summertown, TN: Book Publishing Co., 1985.

McDougall, Walter A. *The Heavens and the Earth*. New York: Basic Books, 1985.

Maddox, Brenda. *Beyond Babel*. New York: Simon & Schuster, 1972.

Magdelenat, Jean-Louis. "The Controversy over Remote Sensing." In *People in Space*, edited by James Katz. New Brunswick, NJ: Transaction Books, 1985. Pp. 129, 130–31.

Mausel, Paul, and Guernsey, Lee. "Application of Remote Sensing in Regional Planning." In *Introduction to Remote Sensing of the Environment*, 2d ed., edited by Benjamin Richason. Dubuque, IO: Kendall-Hunt, 1983. Pp. 423–61.

National Commission on Libraries and Information Science. *To Preserve the Sense of Earth From Space.* Washington, DC: General Printing Office, 1984.

The 1986 Satellite Directory. Potomac, MD: Phillips Publishing Inc., 1986.

Office of Technology Assessment. *Remote Sensing and the Private Sector.* Washington, DC: Government Printing Office, 1984.

Pelton, Joseph. "The Proliferation of Communications Satellites: Gold Rush in the Clarke Orbit." In *People in Space,* edited by James Katz. Dubuque, IO: Transaction Books, 1985. Pp. 98, 105–6.

Pelton, Joseph, and Filep, Robert. "Tele-Education by Satellite." In *Toward International Tele-Education,* edited by Wilbur Blume and Paul Schneller. Boulder, CO: Westview, 1984. Pp. 149–88.

Plowman, Edward W. *Space, Earth, and Communication.* Westport, CT: Greenwood Press, 1984.

Poole, Ithiel de Sola. *Technologies of Freedom.* Cambridge: Harvard University Press, 1983.

Richason, Benjamin. "Landsat Platforms, Systems, Images, and Image Interpretation." In *Introduction to Remote Sensing of the Environment,* 2d ed., edited by Benjamin Richason. Dubuque, IO: Kendall-Hunt, 1983.

Rundquist, Donald, and Samson, Scott. "Application of Remote Sensing in Agricultural Analysis." In *Introduction to Remote Sensing of the Environment,* 2d ed., edited by Benjamin Richason. Dubuque, IO: Kendall-Hunt, 1983. Pp. 317, 328–30.

Sabins, Floyd. *Remote Sensing: Principles and Interpretation.* San Francisco: W. H. Freeman, 1978.

Schichtle, Cass. *The National Space Program: From the Fifties into the Eighties.* Washington, DC: National Defense University, 1983.

Schiller, Dan. *Telematics and Government.* Norwood, NJ: Ablex Publishing, 1982.

Schnapf, Abraham. *Communications Satellites: Overview and Options for Broadcasters.* New York: American Management Association, 1982.

Simon, Samuel. *After Divestiture.* White Plains, NY: Knowledge Industry Publications, 1985.

Singleton, Loy. *Telecommunications in the Information Age.* Cambridge, MA: Ballinger Publishing Co., 1983.

Smith, Delbert. *Communications via Satellite: A Vision in Retrospect.* Boulder, CO: Westview, 1976.

Smith, Marcia. "U.S. International Space Activities." In *People in Space,* edited by James Katz. New Brunswick, NJ: Transaction Books, 1985. Pp. 82, 87.

Stares, Paul. *The Militarization of Space: U.S. Policy, 1945–84.* Ithaca, NY: Cornell University Press, 1985. Pp. 14–15.

Taylor, Thomas. "Telecommunications and Public Safety." In *Telecommunications and Productivity,* edited by Mitchell Moss. Reading, MA: Addison-Wesley, 1981.

Turnill, Reginald. *Jane's Spaceflight Directory.* London: Jane's Publishing Co., 1984.

Villmow, Jack. "Application of Remote Sensing in Weather Analysis." In *Introduction to Remote Sensing of the Environment,* 2d ed., edited by Benjamin Richason. Dubuque, IO: Kendall-Hunt, 1983. Pp. 476–97.

PERIODICALS

"Aerospace Spotlight." *Aerospace America* 24 (October 1986): 1.

Anderson, John. "Remote Sensing Finds Down-to-Earth Applications." *Commercial Space* 1 (Fall 1985): 70, 73.

Andrews, David. "The Legal Challenge Posed by the New Technologies." *Jurimetrics* 24 (Fall 1983): 43.

Anglin, Richard. "Mobile Satellites: The New Business in Space." *International Space Business Review* 1 (June/July 1985): 6–15.

"Ariane V16 Launch Provides Bright Spot in Recent Space Turmoil." *Space Commerce Bulletin*, 28 February 1986, 7–8.

"Ariane Takes Advantage of Shuttle Halt to Rationale Pricing With Increase." *Space Commerce Bulletin*, 23 May 1986, 8.

"AT&T Buying Ford Aerospace Satellite Operating Subsidiary." *Space Commerce Bulletin*, 19 December 1986, 3–4.

Barnett, Hayward. "Spreading the Word." *Wire Communications*, September 1985, 43.

Bassler, John, and Carter, William. "Geodesy Goes High Tech." *Aerospace America* 24 (April 1986): 72.

Boudreau, P.M., and Breithaupt, R.W. "Canadian MSat Program Moves Out." *Aerospace America* 23 (June 1985): 62.

Bourely, Michel. "The Contributions Made by International Organizations to the Formation of Space Law." *Journal of Space Law* 10 (1982): 139.

Brauer, Frank. "Secure UHF Satellite Links for the '90s." *Aerospace America* 23 (June 1985): 54–57.

Brodsky, Robert; Wolfe, Malcolm; and Pryke, Ian. "Foreign Launch Competition Growing." *Aerospace America* 24 (July 1986): 7.

Budiansky, S. "Business in Space Gets a Boost." *US News & World Report* September 1986, 60–61.

Bullock, Chris. "Ariane 4 and Its Competitors." *Interavia* 5 (1986): 551.

Burnett, James. "Commercial Remote Sensing Data Policy." *American Bar Association Forum Committee on Air and Space Law*, 24 February 1984.

"Business." *Space Commerce Bulletin*, 24 October 1986, 7.

"Business." *Space Commerce Bulletin*, 7 November 1986, 7.

Campagna, Richard. "Video and Satellite Transmission Piracy in Latin America: A Survey of Problems, Legal Strategies, and Remedies." *International Lawyer* 20 (Summer 1986): 961.

Caplan, Lawrence. "The Case For and Against Private International Communications Satellite Systems." *Jurimetrics* 26 (1986): 198.

Chapman, John, and Warren, Gabriel. "Direct Broadcast Satellites: The ITU, UN, and the Real World." *Annals of Air and Space Law* 4 (1979): 413.

Clarke, Arthur C. "Extra-Terrestrial Relays." *Wireless World* 51 (October 1984): 305.

Clayburgh, William. "On Buying Spaceships, Profits, and Dreams." *International Space Business Review* 1 (July/August 1986): 55.

Coleman, Herbert. "Space Insurance." *Commercial Space* 2 (Fall 1985): 61.

Colino, Richard. "A Chronicle of Policy and Procedure: The Formulation of the Reagan Administration Policy on International Satellite Telecommunications." *Journal of Space Law* 13 (1985): 103.

Colino, Richard. "Intelsat: Facing the Challenge of Tomorrow." *Columbia Journal of International Affairs* 39 (1985): 129.

"Combining Metasat, Landsat Sensors on Polar Orbiter May Avoid Budget Cuts." *Space Commerce Bulletin*, 4 April 1986, 3.

"Commerce Department Officials Explain Landsat Cutback to Hill Staffers." *Space Commerce Bulletin*, 2 January 1987, 6.

Condoin, Pierre. "Brazil Aims for Self-Sufficiency in Space." *Interavia* 5 (1986): 99–101.

Covault, Craig. "Economic Competition." *Commercial Space* 1 (Fall 1985): 18–21.

Dougherty, Craig. "A Financial Prospective of Satellite Projects." *International Space Business Review* 1 (June/July 1985): 64.

De Lavan, Joanne. "HP's High Tech Net." *Satellite Communications* 9 (March 1985): 32.

"EOSAT May Scale Back or Halt Satellite Development Plans." *Space Commerce Bulletin*, 21 November 1986, 4.

Evans, L.J. "National Commercial Space Initiatives: Legal Implications." *American Bar Association Forum Committee on Air and Space Law*, 1 November 1984.

"FCC Opens the Skies to DBS." *Broadcasting*, 28 June 1982, 27.

Feldman, Nathaniel. "Aerospace Mobile Communications: Building the Mass Market." *Aerospace America* 23 (June 1985): 50.

Gawdun, Michael. "Virtual Private Networks." *Telecommunications* 20 (April 1986): 59–61.

Ghais, Ahmed, et al. "Broadening Inmarsat Services." *Aerospace America* 23 (June 1985): 66.

Glaser, Peter, and Brender, Mark. "The First Amendment in Space: News Gathering From Satellites." *Issues in Science and Technology* 3 (Fall 1986): 60–67.

Godwin, Kimberly. "The Proposed Orion and ISI Transatlantic Satellite Systems: A Challenge to the Status Quo." *Jurimetrics* 24 (1984): 297.

Goldberg, Donald. "Captain Midnight, HBO, and World War III." *Mother Jones* 11 (October 1986): 26–29, 48–53.

Greeley, Brendan. "Commercial Marketing of Landsat Data Begins." *Commercial Space* 1 (Fall 1985): 66, 68.

Greenberg, Joel, and Gaelick, Carole. "Simulation to the Rescue of Space Insurance." *Aerospace America* 24 (August 1986): 62.

Hashemi, Michael. "The Replacement of the World's Largest Private Network." *Telecommunications,* April 1986.

"House, Air Force Pursue Separate Paths to Encourage Commercial ELV Market." *Space Commerce Bulletin,* 15 August 1986, 3.

"Indonesia May Buy Back Rescued Satellite." *Space Commerce Bulletin,* 18 July 1986, 6.

"Inmarsat Receives Launch Proposal from Soviets." *Space Commerce Bulletin,* 4 July 1986, 5–6.

"Intec Sets up Committee to Consider Private Launch Insurance Solutions." *Space Commerce Bulletin,* 15 August 1986, 6.

Ito, Youichi. "Recent Trends in Telecommunications Regulation and Markets in Japan." *Jurimetrics* 25 (1984): 70.

Jakhu, Ram. "The Legal Status of the Geostationary Orbit." *Annals of Air and Space Law* 7 (1982): 333, 349–50.

Jaques, Bob. "Canada's Multi-Role Radar Satellite." *Interavia* 5 (April 1986): 433–34.

Jefferson, Pat. "Congress Delves into ELV's." *Aerospace America* 24 (September 1986): 9, 13.

Kane, Francis, and Scheerer, John. "The Global Role of Navstar." *Aerospace America,* July 1984, 42.

Latin, Howard; Tannehill, Gary; and White, Robert. "Remote Sensing Evidence and Environmental Law." *California Law Review* 64 (1976): 1300.

"Launchers." *Space Commerce Bulletin,* 9 May 1986, 9.

Leive, David. "Intelsat In a Changing Telecommunications Environment." *Jurimetrics* 25 (1984): 82.

Lerner, Eric. "Designing Communications Satellites: Intelsat VI and Aussat." *Aerospace America* 23 (May 1985): 93.

Lerner, Eric. "Space Insurance: Who Takes the Risk?" *Aerospace America* 24 (March 1986): 3, 12.

Lingl, R. "Risk Allocation in International Interbank Electronic Fund Transfers: CHIPS & SWIFT." *Harvard International Law Journal* 22 (1981): 621.

Logue, Timothy. "US Decisions on Pacific Telecommunications Facilities: Letting a Million Circuits Bloom?" *Jurimetrics* 27 (1986): 65.

Marks, Herbert. "Regulation and Deregulation in the United States and Other Countries." *Jurimetrics* 25 (1984): 5.

McKenna, Julianne. "Bypassing Intelsat: Fair Competition or Violation of the Intelsat Agreement." *Fordham Journal of International Law* 8 (1985): 479.

Miller, Jonathan. "Gold Rush in the Sky." *Channels 1986 Field Guide,* (1986): 53.

Mordoff, Keith. "Expendable Launchers." *Commercial Space* 1 (Fall 1985): 29–30.

Muiller, Helmut. "Intelsat Goes Commercial." *Interavia* 5 (February 1986): 195.

Myers, James. "Terminating Business Use of the Shuttle." *International Space Business Review* 2 (July/August 1986): 29.

Myers, James. "Federal Government Regulation of Commercial Operations Using Expendable Launch Vehicles." *Journal of Space Law* 12 (1984): 40.

"NASA Halts Delta Commercialization Talks with TCI; Allies Are Lobbied for Aid." *Space Commerce Bulletin,* 24 October 1986, 3–4.

"NASA Would Buy Commercial Launch Services." *Space Commerce Bulletin,* 15 August 1986, 3.

Ott, James. "Instant Delivery." *Commercial Space* 1 (Winter 1986): 66.

Owen, David. "Satellite Television." *Atlantic,* June 1985, 45.

Owen, Kenneth. "Inmarsat Takes to the Air." *Aerospace America* 24 (July 1986): 14.

"Panamsat Wins Intelsat Board Approval for Separate System." *Space Commerce Bulletin,* 7 November 1986, 5–6.

Pritchard, Wilbur, and Nelson, Robert. "Challenger Electrifies Transponder Market." *Aerospace America* 24 (May 1986): 9.

Rasoon, S.I. "Predicting Earth's Dynamic Changes." *Aerospace America* 24 (January 1986): 78–82.

"Reaction to White House Space Policy Mixed." *Space Commerce Bulletin,* 29 August 1986, 2.

"Remote Sensing Satellites Can Be Useful News Tool, RTNDA Is Told." *Space Commerce Bulletin,* 20 June 1986, 6–7.

Roland, Alex. "Triumph or Turkey." *Discover* 6 (November 1985): 29.

Rovell, Robert, and Cuccia, Louis. "A New Wave of Communications Satellites." *Aerospace America* 22 (March 1984): 43.

Rutowski, A.M. "The Integrated Services Digital Network: Issues and Options for the Future." *Jurimetrics* 24 (Fall 1983): 19.

Sanchez, Ernest, and Mooney, Diane. "Current Legal Developments in Direct Broadcasting Satellites." *Jurimetrics* 25 (Winter 1985): 218.

Sarreals, Cheryl. "International Telecommunications Satellite Services: The Spirit of Cooperation vs. the Battle for Competition." *Jurimetrics* 26 (1986): 267.

Satellite Week, 11/10/86, at p. 1.

"Senate Appropriations Committee Backs Use of DOD Money for Shuttle." *Space Commerce Bulletin,* 26 September 1986, 3.

"Senate Authorizes Landsat Funding." *Space Commerce Bulletin,* 15 August 1986, 7.

"Shuttle Inquiry Is Forging Ahead but Delay Seems Likely." *Space Commerce Bulletin,* 14 February 1986, 2–3.

"Shuttle Will Refly for Half Price." *Space Business News,* 21 October 1985, 1.

Simonoff, Jerome. "The Financial Implications of the Insurance Situation." *International Space Business Review* 1 (June/July 1985): 62.

Slate, Edward, and Popko, John. "The Next Five Years in Communications." *Telecommunications* 20 (January 1986): 49.

Smith, Ilene. "Fiber Optics: Can Satellites Compete?" *Satellite Communications,* February 1985, 22.

Smith, Ilene. "Hello, Federal." *Satellite Communications* 9 (July 1985): 27.

Space Calendar, 10 October/20 October 1985, 3.

Space Calendar, 2 June/8 June 1986, 3–4.

Space Calendar, 29 September/5 October 1986, 4.

Space Calendar, 24 November/30 November 1986, 4.

"Space Commerce Notebook." *Space Commerce Bulletin,* 26 September 1986, 9.

"Space Industries and Westinghouse Form Partnership to Develop Spacecraft." *Space Commerce Bulletin,* 10 October 1986, 2.

Staple, Gregory. "The New World Satellite Order: A Report from Geneva." *American Journal of International Law* 80 (1986): 699.

Szasz, Paul. "Report on the 25th Session of the Legal Sub-Committee of the UN Committee on the Peaceful Uses of Outer Space, 24 March–11 April 1986." *Journal of Space Law* 14 (1986): 48.

NEWSPAPERS

"Air Force Blows Up a Test Missile as Flight Over Pacific Goes Awry." *New York Times*, 29 August 1986, A6.

Berg, Eric. "British Telecom and AT&T Venture Expected." *New York Times*, 22 July 1985, D21.

Blakeslee, Sandra. "Satellites Assist Pacific Fishermen." *New York Times*, 18 December 1983, I40.

Boffey, Philip. "7-Year Schedule Is Set in Space Shuttle Tasks." *New York Times*, 4 October 1986, 5.

Boyd, Gerald, "Aides Say Reagan Will Order NASA to Build Shuttle." *New York Times*, 15 August 1986, A1.

Boyer, Peter. "HBO Piracy Incident Stuns Other Satellite Users." *New York Times*, 29 April 1986.

Bramford, James. "America's Supersecret Eyes in Space." *New York Times*, 13 January 1985, VI38, 39.

Broad, William. "Civilians Use Satellite Photos for Spying on Soviet Military." *New York Times*, 7 April 1986, A1.

Broad, William. "Military Missions to Dominate Role of Space Shuttle." *New York Times*, 24 March 1986, A1.

Broad, William. "Newest Titan Groomed as Rival for Shuttle." *New York Times*, 20 May 1986, C1.

Broad, William. "Satellite Photos Appear to Show Construction of Soviet Space Shuttle Base." *New York Times*, 25 August 1986, A20.

Broad, William. "Titan Loss May Force Early Use of Shuttle." *New York Times*, 22 April 1986, C1.

Buckley, William F. "Precision Sailing." *New York Times*, 19 May 1985, VI134, 138.

Burnham, David. "Loophole in Law Raises Concern About Privacy in Computer Age." *New York Times*, 19 December 1983, A1.

Burns, Joseph. "China's Proud Space Program." *New York Times*, 19 May 1986, D10.

Feder, Barnaby. "MCI Struggles to Ward off Deregulation's Sting." *New York Times*, 13 August 1986, III6.

Gleick, James. "Errant Rocket Destroyed by Ground Control." *New York Times*, 28 August 1986, B11.

Gleick, James. "The Future of Long-Term Broadcasting Is Bright." *New York Times*, 20 October 1985, IV7.

Greenhouse, Linda. "The Wiretapping Law Needs Some Renovation." *New York Times*, 1 June 1986, IV4.

Manber, Jeffrey. "Insurers Shy Away From Satellites." *New York Times*, 19 January 1986, VI17.

McGill, Douglas. "Teleport on Staten Island Envisioned as City's Link to a Bright Future." *New York Times*, 10 September 1983, 25.

Miller, Judith. "Latest Failure Cripples West's Satellite Ability." *New York Times*, 1 June 1986, A1.

Mohr, Charles. "Pentagon Fears Delays on Future Spy Satellites." *New York Times*, 24 February 1986, B6.

Molotsky, Irvin. "Chernobyl and the 'Global Village.'" *New York Times*, 8 May 1986, A22.

Olushman, Jahn. "Air Force to Build New Rocket Fleet to Loft Satellites." *New York Times*, 1 August 1986, A1.

Pace, Eric. "TV Curb Backed by UN Assembly." *New York Times*, 11 December 1982, 6.

Peterson, Ivor. "Scrambling of Signals Today Thwarts TV Dish Antennas." *New York Times*, 15 January 1986, A1.

Pollack, Andrew. "Plan for TV by Satellite Falls Apart Over Risks." *New York Times*, 1 December 1984, 31.

Sanger, David. "Commercial Launching Shifted to Private Sector." *New York Times*, 16 August 1986, 1.

Sanger, David. "Comsat, Contel Plan to Merge." *New York Times*, 30 September 1986, D1.

Sanger, David. "Japan Missing a Chance to Be a Leader In Space." *New York Times*, 8 December 1986, D1.

Sanger, David. "NASA Will Launch Earth Satellites Using its Rockets." *New York Times*, 12 March 1986, A1.

Sanger, David. "Phone Group Plans Pacific Cable Link." *New York Times*, 24 February 1986, D4.

"Satellites Outpace Customers." *New York Times*, 10 April 1984, D1.

Schmitt, Eric. "Network Keeps Alaska Legislators in Touch." *New York Times*, 20 August 1985, A21.

Sims, Calvin. "Federal Express to End Electronic Mail Service." *New York Times*, 30 September 1986, D1.

Smith, Sally. "Two-Way Cable TV Falters." *New York Times*, 28 March 1984, C25.

"Space Shuttle and Commerce." *New York Times*, 11 March 1986, D2.

Stuart, Reginald. "Intelsat Proves a Tough Foe." *New York Times*, 14 July 1986, D2.

Stuart, Reginald. "TV Groups Reach Pact on Must-Carry Rule." *New York Times*, 28 February 1986, C23.

Stuart, Reginald. "U.S. Limits Local Rules for Satellite Receivers." *New York Times*, 15 January 1986, A16.

Sullivan, Joseph. "Cordless Phones Raise an Eavesdropping Issue." *New York Times*, 11 March 1986, B3.

Sullivan, Walter. "Distress Satellite Loses Orbit." *New York Times*, 14 July 1984, 46.

Sullivan, Walter. "New Satellite Technology Will Beam TV to Remote Regions." *New York Times*, 30 August 1983, C3.

Toner, Robin. "Using Satellites to Reach Voters in the Out There." *New York Times*, 16 December 1985, B10.

"U.S. Signs a Satellite Pact with 3 Countries." *New York Times*, 18 October 1984, A18.

"Weather Satellite Launched Atop Rebuilt Rocket." *New York Times*, 18 September 1986, A17.

Weisman, Steven. "India Carrics its Dreams into Space." *New York Times*, 23 December 1986, C1.

Wilford, John Noble. "Plans for Space Shuttles Flourish Outside the US." *New York Times*, 12 August 1986, C1.

Wilford, John Noble. "Spacecraft Detects Sahara's Past." *New York Times*, 26 November 1982, A1.

Wilford, John Noble. "Test of European Rocket to Delay Launching." *New York Times*, 3 December 1986, A18.

Wilford, John Noble. "Weather Satellite Marks a Birthday." *New York Times*, 2 April 1985, C2.

BIBLIOGRAPHY

GOVERNMENT PUBLICATIONS

Commercial Space Launch Act, PL 98-575, 49 USC 2601 (1984).

Office of the Secretary of Transportation, United States Department of Transportation, "Commercial Space Transportation: Licensing Process for Commercial Space Launch Activities," OST Docket 42885, Notice 85-3 (2/15/85).

Communications Satellite Act of 1962, PL 87-624, 76 Stat. 419 (1962).

National Aeronautics and Space Act of 1958, PL 85-568, 72 Stat. 426 (1958).

Internal Revenue Code, SEC. 48 (a) (2), (7).

Executive Order No. 12465, 49 Fed. Reg. 7211 (1984).

Presidential Directive/National Security Council NSC-42, July 4, 1982.

Pub. L. 73-416, US Statutes at Large (1934).

Pub. L. 98-166, HR 3222 (1983).

Stowe v. *DeVoy*, 588 F. 2d 336 (2d Cir. 1978).

U.S. v. *Cotroni*, 527 F. 2d 708 (2d Cir. 1975).

70 FCC 2d 1460, Docket 78-374 (1979).

90 FCC 2d 1159, Docket 80-634 (1982).

Treaty on Principles Governing the Activities of States in the Exploration and Use of Outer Space, Including the Moon and Other Celestial Bodies, 18 UST 2410, TIAS 6347 (1967).

The Agreement on the Rescue of Astronauts, the Return of Astronauts and the Return of Objects Launched into Outer Space (Return and Rescue Treaty) 19 UST 7570, TIAS 6599.

The Convention on International Liability for Damage Caused by Space Objects (Liability Treaty) 24 UST 2389, TIAS 7762.

The Convention on Registration of Objects Launched into Outer Space (Registration Treaty) 28 UST 695, TIAS 8480.

Index

expendable launch vehicles (ELV's),
73–76, 119–21, 124, 139, 140;
see also launch services; space
shuttle
Explorer I, 7

F

fascsimile transmission, 23–24, 31
FCC (Federal Communications
Commission), 10, 76–77, 114
Federal Express Zap Mail, 24, 31,
94
fiberoptics, 28–29, 44, 140, 142,
144
Financial Satellite (FinanSat), 94,
125
First Interstate Bancorp, 11
footprints, broadcasting, 14, 115,
133

G

Galaxynet, 12
Gavkosmos, 90
General Agreement on Trades and
Tariffs, 112
General Dynamics, 89
General Electric, 86
geodesy, 59–60
Geosat, 58
Geostar, 63, 100
geostationary orbit, 4–5, 9, 15–16,
51, 129–31, 145
Glenn, John, 8
Globesat, 94
GOES satellite, 86, 124
Gramm-Rudman Act, 56
Great Wall Industrial Corporation,
90
GTE, 92

H

Hermes shuttle, 91
HBO (Home Box Office), 39, 132–
33
hotel broadcasting networks, 48

Hughes Aircraft Co., 85–86, 92, 94
hydrology, 61

I

IBM, 11, 92
Inmarset (International Maritime
Satellite Organization), 11, 28,
82–83, 90, 92, 93
Insat-2, 100
integration of communication
services, 108–10
Intelsat (International
Telecommunications Satellite
Consortium), 9, 11, 12, 22, 78,
79, 81–83, 92, 94, 95, 125–27,
142, 144
Intelsat Business Services (IBS), 82
interference, 107–8, 115
International Civil Aviation
Organization (ICAO), 84
International Council of Scientific
Unions, 6
International Frequency
Registration Board (IFRB), 83
International Geophysical Year, 6
International Satellite Inc. (ISI), 94
International Telecommunications
Union. *See* ITU
Intersputnik, 83, 93
Interstate Commerce Commission
(ICC), 76
issues for operators, 119–35
economic, 121–23
in many-to-one information
transfer, 133–35
in one-to-many information
transfer, 129–33
in one-to-one information
transfer, 125–29
political, 119–21
technological, 123–25
issues for users, 103–17
economic, 104–5
in many-to-one information
transfer, 116–17